Born This Wa

Becoming, Being, and Understand

Part 2: The Practice of Science and the Rise of Molecular Biology

Richard A Lockshin, Ph.D.

Dedicated to:

Zahra Zakeri

Who makes every day more exciting and more enjoyable than the last.

Published by Richard A. Lockshin

Smashwords Edition (Electronic)

Printed Edition available through

http://www.amazon.com

Discover other titles by Richard A. Lockshin at Smashwords.com
(*Born This Way: Part 1: The Origins of Modern Biological Science*, by Richard A. Lockshin; other titles through Amazon).

Cover Photo courtesy of Igor Siwanowicz,
http://photo.net/photodb/user?user_id=1783374, reprinted with permission

Contents

PREFACE: SCIENCE AND THE SINGLE HUMAN

How can we be both science-exacting and science-ignorant?

We run our lives by science. We expect our medicines, our physicians, the robots that we use for surgery, our food and our water, to be 100% free from harm, and we sue should we find that they are not perfect. We depend absolutely on our GPS systems and on the physics by which machines as big as a 20-story building carry us seven miles above the surface of the earth, half way around the earth. We rely on our meteorologists to predict bad weather and we fire politicians who do not address the storms. We revel in growing varieties of flowers and food plants in many different environments. We demand seedless fruits, vegetables in multiple colors, weed-free lawns, and insect-resistant flowers with huge blossoms with many layers of petals. We marvel that computer-guided weapons can destroy property and other people without putting American lives in danger. We clamor for the latest wonder drug, including "designer drugs" tailored to the specific disease and genetics of the patient, and we adore the new breeds of dogs and cats. We subject ourselves to chemical and surgical body-enhancing procedures, and athletes seek out hormones and growth factors that will enhance their abilities. Many of us rapidly adapt to the latest scientific argument, eagerly buying products that are claimed to suppress "free radicals," enhance "metabolism," or remove "toxins". We argue vehemently about the use of "fossil fuels" and the different means of extracting them, as opposed to nuclear energy and solar energy.

And yet many of us deny science.

Half of us do not believe that humans evolved from similar species over periods counted in millions of years.

Some think that the earth is flat.

A majority do not accept all the documentation that the earth is warmer than 100 years ago, consider that humans have not affected the climate, or do not accept the predictions of the biological and geological consequences of the changing of global temperature.

Some of this comes about because science appears so foreign--hermetic, arcane, filled with complex calculations, torture of animals, unpleasant and disagreeable chemicals, impossible words, and basically incomprehensible babble. No music, no songs, no touchdowns or home runs, in short, the heart of boring geekdom.

And yet...

Science is within reach of everyone. Everyone, in one way or another, is a scientist, asking questions, wanting to know how things work, exploring relationships of cause and effect. Scientists themselves enjoy life. We enjoy dancing, singing, parties, sports. Most importantly, we can explain what we are doing, and why we do it. We work on the assumption that we can best understand cause and effect by using our logic and our senses to test our hypotheses. It is something that everyone does in one form or another. Sometimes our rules are a little different. What we ask is that we, scientists and lay, explain to each other what we understand to be the rules.

So what is it like to be a scientist? Think ten-year old at a magic show, invited onstage to see how the magician performs his tricks. Alternatively, we all try to understand "the system". This must have been a priority for the earliest hunter-gatherers: "If the antelope came to the waterhole last evening, it is likely to come back this evening." Lottery players rely heavily on working out ways to beat "the system", coming up with complicated formulae for choosing numbers. Even in our everyday chicanery (how to find a parking meter with unexpired time) we take pleasure in outsmarting "the system". This is how a lab scientist spends his or her days, trying to find ways in which we can trick nature into revealing her secrets. To find a new secret, to reveal a new mechanism through which an organism accomplishes its goals, is a source of enormous pleasure. We spend our days matching our wits with the most imaginative resources that God or evolution could devise.

Of course, it is also possible to consider the work of a scientist to be filled with excruciating boredom or frustration. The truly satisfying or elegant (see next paragraph) experiment is a rare bird indeed. Most of our time is spent reading about what others have found and considering the options and permutations of those results, or in trying to improve experiments. Experiments usually do not work (at least in the biomedical sciences; for large-scale physics experiments, the cost is so high that all limitations have to be identified well in advance). In the biomedical sciences, the initial results of an experiment are often ambiguous or not clean enough for publication, and we invest considerable effort in asking questions like, "It is possible for the experimental value to be too low but not for it to be too high. It is also possible for the control value to be too high. What can I do to make sure that the control value is not artifactually too high, so that I can reveal the difference?" And we spend days trying minor variants of the same experiment, tweaking the system to make the results a little better. As I often tell students who aspire to a career in research, the personality of a researcher is that he or she must simultaneously abhor and tolerate ambiguity. One has to tolerate it, because initial experiments are often not clear enough to come to a precise conclusion. On the other hand, one has to abhor ambiguity enough to keep going, to be committed to resolving it.

At the end of the day, one strives to carry out an "elegant experiment". Elegance is highly prized among researchers. The term connotes an experiment so pure and clean, so efficient, that there can be only one conclusion: *Quid Erat Demonstrum*. Einstein's eclipse experiment (p. 46) was elegant. In a similar fashion, Karl Landsteiner's prediction of the shape of an antibody molecule could be considered to be elegant because his interpretation of the meanings of the properties of the molecule proved to be correct. Briefly, he considered that, when antigen (the protein that reacts with an antibody) was added to a constant amount of antibody, when there was too little antigen, there would be no precipitate. When the amount of antigen relative to antibody was approximately 1:1, there would be a precipitate; but when the amount of antigen exceeded the amount of antibody, there would be no precipitate or a previously-formed precipitate would dissolve. He concluded that the antibody must have two binding sites, allowing it to form chains or polymers when the ratio of concentrations was correct (Fig. p1a). When the structure of antibody molecules was finally elucidated, about 50 years later, it was truly gratifying to see the prediction verified (Fig. p1b). (Landsteiner also identified the major blood groups—Rh factor, A, B, and O—for which he was awarded the Nobel Prize.) Other elegant experiments would include Lederberg's demonstration (p. 104) of bacterial sex, Watson and Crick's elucidation of the structure of DNA (p. 90), and Crick's demonstration that the genetic code was linear (p. 100).

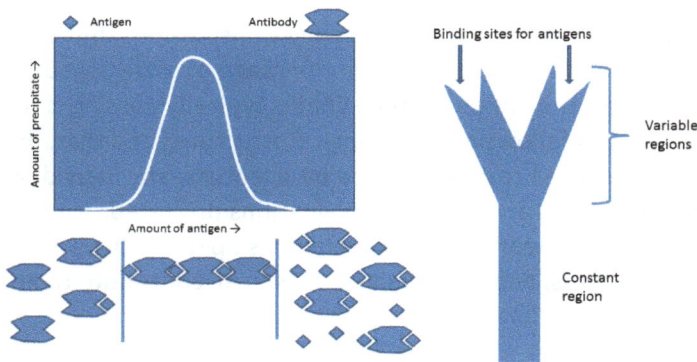

Figure p1a (left). Landsteiner's prediction. The antibody can link to the antigen through two points. When the ratio is approximately 1:1, it can form insoluble chains of antigen-antibody-antigen-antibody-antigen... as in center of graph. If there is more antibody than antigen (left of graph), each antibody molecule can bind only one antigen, and the chain will not form. If there is more antigen than antibody, each antibody molecule will be saturated with extra antigen, and the chain will not form. Figure p1b (right): The basic structure of an antibody molecule (here highly simplified). It proved to be a two-headed molecule, as predicted. One region is constant, while the heads, to which antigens bind, come in many forms, allowing us to make antibodies that can identify many different types of antigens. The whole story of how this is done is fascinating and worthy of retelling at length.

Elegance is recognized in most walks of life. For instance, at one point in my life I was writing computer code. I would write a program that would do what I wanted, but it had some workarounds and patches (what computer people call "kluges") because I didn't know how to address one issue. In nearly addictive fashion, I would work (unnecessarily) on the program until it was lean, linear, and flawless. We all do this in one way or another: fold shirts perfectly, arrange a bookshelf in a specific order, make sure the silverware is perfectly aligned on the table, confirm that there is not a stray bubble in a painted object or a fold in a caulk line… It is part of demonstrating mastery of "the system", demonstrating that one understands the workings so as to be in perfect control of them. Such are the sentiments of researchers.

And what about those who are not scientists? Well, that would be a very small proportion of humankind, if any exist at all. We are all scientists when we query how things work, and we are all so hugely affected by the impacts of our curiosity that we cannot avoid, as we use the artifacts of our civilization that derived from scientific ideas (computers, electricity, to name the most complex, but also including virtually every crop or domesticated animal, none of which remotely resemble their wild forms), or we contemplate headlines every day: to vaccinate or not; use genetically modified organisms (GMO) as food or not; the extent to which climate change is real and will affect us; what types of fuel to use for energy; the effect of nerve gases; the science and morality of fertility and the beginning of life; the science and morality of stem cell research; the extent to which the end of life should be regulated; mechanical and laboratory-grown replacement organs; the value of space exploration…

Many of these questions involve highly technical discussions, but the basic premises and assumptions are well within the range of understanding of any reasonably educated and curious adult. What is typically missing is recognition that scientific analysis follows common logical structures and that, by accepting a few rules and approaches commonly used by scientists, any individual can grasp the gist of the arguments and assess the valuations proffered. Thus here I attempt to describe how science works and how scientists operate. Once we get beyond the slogans, shibboleths, and propaganda, we can all contribute to and guide that great human enterprise that is driven by curiosity.

*****

SCIENCE AND THE SINGLE SPECIES

"Science knows no country, because knowledge belongs to humanity, and is the torch which illuminates the world. Science is the highest personification of the nation because that nation will remain the first which carries the furthest the works of thought and intelligence." (Louis Pasteur)

"Consider again that dot. That's here. That's home. That's us. On it everyone you love, everyone you know, everyone you ever heard of, every human being who ever was, lived out their lives. The aggregate of our joy and suffering, thousands of confident religions, ideologies, and economic doctrines, every hunter and forager, every hero and coward, every creator and destroyer of civilization, every king and peasant, every young couple in love, every mother and father, hopeful child, inventor and explorer, every teacher of morals, every corrupt politician, every "superstar," every "supreme leader," every saint and sinner in the history of our species lived there – on a mote of dust suspended in a sunbeam.

The Earth is a very small stage in a vast cosmic arena. Think of the rivers of blood spilled by all those generals and emperors so that in glory and triumph they could become the momentary masters of a fraction of a dot. Think of the endless cruelties visited by the inhabitants of one corner of this pixel on the scarcely distinguishable inhabitants of some other corner. How frequent their misunderstandings, how eager they are to kill one another, how fervent their hatreds. Our posturings, our imagined self-importance, the delusion that we have some privileged position in the universe, are challenged by this point of pale light. Our planet is a lonely speck in the great enveloping cosmic dark. In our obscurity – in all this vastness – there is no hint that help will come from elsewhere to save us from ourselves.

The Earth is the only world known, so far, to harbor life. There is nowhere else, at least in the near future, to which our species could migrate. Visit, yes. Settle, not yet. Like it or not, for the moment, the Earth is where we make our stand. It has been said that astronomy is a humbling and character-building experience. There is perhaps no better demonstration of the folly of human conceits than this distant image of our tiny world. To me, it underscores our responsibility to deal more kindly with one another and to preserve and cherish the pale blue dot, the only home we've ever known."

—Carl Sagan, Pale Blue Dot: A Vision of the Human Future in Space, 1997 reprint, pp. xv–xvi

~~~~~

Chapter 1: Why can't scientists speak English?

Conversational language is easy and pleasant, but it demands a great deal of interpolation from the interlocutors. Statements are generally not clear, sentences are incomplete, and terms are casual. Witness the classic "Honey, would you get that for me?" in which the "that" refers to something talked about a few minutes before, pointed to, or commonly used in the past and generally intuited based on long experience with one's partner. Or: "That's cool, man!" meaning something presumptively agreeable, but specifically what is agreeable and how it is agreeable are left to the listener to recognize and appreciate. Obscenities are frequently used for emphasis, shortcutting a more specific description of the uniqueness or surprising nature of the subject of reference. Consider how often one or two particular obscenities are used to replace numerous possible adjectives, modifiers, and adverbs.

The heart of science is its ability to test and replicate or refute another's claim. Thus we must understand exactly what the claim is and vagueness will not do. We have to know exactly what was done, so that we can consider if anything was missed or overlooked. Details are important. The comment, "A really gorgeous girl" (a.k.a. "She's f*in hot!") would conjure different images to different people. Police, for instance, would expect indications of age, height, weight, skin and hair color, type of coiffure, dress, and any other details that would help them to identify a specific individual among many "gorgeous girls". A poet does something quite different.

> "WHENAS in silks my Julia goes,
> Then, then, methinks, how sweetly flows
> That liquefaction of her clothes."

conjures a very pleasant and strikingly effective and evocative image, but the image _evokes_: to each individual that image will be different, depending on sensations, loves, and impressions unique to each individual and that individual's experiences. It is an evoked and therefore imaginative rather than perceived image. The poet Robert Herrick might contemplate a very concrete image but, other than by perhaps a very generic exclusion of some individuals, we cannot identify Julia from the poem. Five observers might identify five different women in a crowd as Julia.

Scientists, like police, need specifics, in this case in order to replicate an experiment. There is a lot of difference between "I mixed these solutions in a beaker" and "I took 50 ml of a 5% (weight/volume) solution of analytical grade sodium hydroxide, held on ice, and to this I added, dropwise, 50 ml of a 1% (volume/volume) solution of concentrated (37%) analytical grade hydrochloric acid in distilled water." We demand details, and we demand precision. As in a jury trial, a car is not "speeding"; it is "going at least 75 miles per hour"; someone seen committing a crime is not "big"; he is "between 6 and 6' 2" tall, and heavyset".

There is another layer to this demand, and that is that scientists speak to each other throughout the country, and throughout the world. As noted in the theme for this book, "**Science knows no country, because knowledge belongs to humanity, and is the torch which illuminates the world.** " Sometimes, there are regional differences in common speech. "Coffee regular" in one part of the country means coffee with cream, and in another part black coffee. If you ask for a "hoagie" in the Southwest, you will be greeted with a look of incomprehension, but you may recognize a sandwich that is locally called a submarine sandwich, and which in New York is called a hero. I even saw once in a Spanish neighborhood that the latter name was spelled "jiro," as it would be if the name were natively Spanish—as opposed to a "gyro," a sandwich made from a mold of pressed meat rotated on a spit, in the Greek fashion. (See a series of maps[1] on the subject.) This would never do in science. Imagine the confusion if a US-based scientist tried to compare notes with a European scientist concerning the turkey[2], a bird found in North America and named because early settlers thought it might be the same bird known in Europe as the turkey-bird, a type of guinea fowl named for the presumption that it came from Turkey (the country). (To add to the confusion, the Turks call the latter bird "hindi," or "bird from India", and the name guinea fowl was given on the assumption that it came from the west coast of Africa. The French name for the American turkey, "dinde" derives from "from the Indies," meaning the West Indies, from the original assumption that Columbus reached India. The French words for guinea pig and horse chestnut follow the same rule.)

Figure 1.1. Left, a guineafowl (*Numida meleagris*), photographed in South Africa. Right, the North American turkey *(Meleagris gallopavo)*. Lack of familiarity with the birds and their origins resulted in confusion of the common names, reflected in the use of the Greek name for guineafowl (*Meleagris*) being used as the genus name for the North American bird.

Similar confusion would exist if one scientist was thinking of the European robin[3] and another was thinking of the North American thrush[4] that colonists mistook for the European bird and therefore called robin. A meerkat[5] (Dutch: lake cat) is not at all related to cats. One could go on and on.

Figure 1.2. Left, a European robin (*Erithacus rubecula*), the true "robin red-breast"). Right: The American robin, *Turdus migratorius,* a type of thrush.

This is why scientists take refuge in scientific names, which are not subject to local variation. These latter are binary names in Latin, rather like "John Smith" except that, like Asian names, the family name is first ("Smith John") so that everyone knows to which animal or plant one is referring.

Another issue is that of originality. Quite often, what scientists describe has never been seen before, cannot be seen or detected without specific equipment, or represents an abstraction based on insights from complex understandings. To differentiate among these, we need very precise words. Thus, once the idea of cells was brand new. In fact, cells are called cells because Robert Hooke saw the spaces left behind after the cells (the living part) of plants died, leaving only the relatively woody cell walls[6] behind.

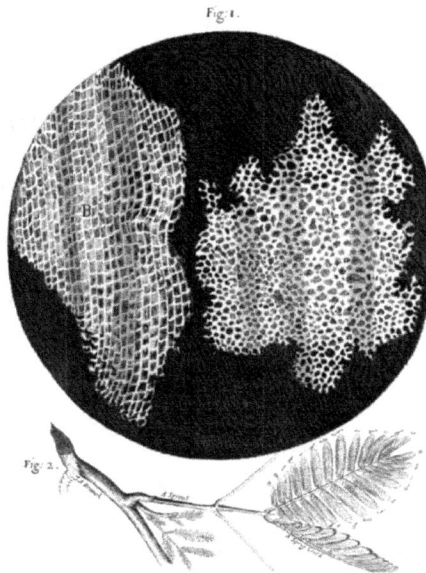

Figure 1.3 Robert Hooke's image of plant "cells," as he saw them through the microscope. What he actually saw was the cell walls left behind after the cells had died. The cell walls give the plant its strength, and the empty spaces serve as conducting tubes to distribute water and nutrients throughout the plant.

The empty little chambers reminded him of monks' cells, giving him the name. Then, when means were developed to fix and preserve the cells in a form that resembled what they might have been in life, Robert Brown confirmed earlier sightings of materials within the cells and defined the nucleus (derived from the word for "nut" and others defined cytoplasm ("the material molded into a container"). Still later, the several organelles were recognized and given individual names based on their shape, content, or presumed function: mitochondria, ribosomes, lysosomes, etc. Today we even have names for the internal components of these organelles (cristae and inner and outer membranes of mitochondria) and different types of some organelles (lysosomes divide into autophagosomes, autophagic vacuoles, primary lysosomes, and others). We even subdivide processes. For instance, we now recognize at least three major ways in which cells die, and many would argue for at least three others.

Each time a scientist introduces a new term, he or she makes an effort to define it in such a way that it will not be confused with a similar but distinct other term, as the term implies a distinct mechanism or function. With the proliferation of knowledge comes the proliferation of terms, and as a matter of efficiency we try to avoid confusion. Ultimately the new term must identify a new process or suggest a new mechanism. If it does not inspire a new way of thinking, it becomes

clutter and eventually will disappear from the literature. As Claude Bernard wrote 150 years ago,

"When one creates a word to characterize a phenomenon, in general everyone understands at this moment the idea that one wishes to have the word express and the exact meaning that one gives to it, but later, by the progress of science, the sense of the word changes for some, while for others the word retains in the language its original meaning. Therefore a discordance results which, often, is such that men, using the same word, express very different ideas. Our language is, in fact, only an approximation, and it is so little precise, even in science, that, if one loses sight of the phenomena to attach oneself to words, one is rapidly outside of reality. One can then only do damage to science when one argues to keep a word that is nothing more than a cause of error, in the sense that it no longer expresses the same idea for everyone."

Finally, we scientists often talk in the common language of our countries, but we use our language in only very restricted ways. For instance, take the word "significant". It comes from "signify," meaning "portend" or "indicate" and derives from the sense of a sign that indicated the outcome of a future event, as a sign—a flock of ravens flying westward—might suggest that the next day's battle would be lost. Today we use the term in many ways: "Unless a significant number of people [= a large number] vote for me.." ; "the newspaper article had a significant [= substantial] impact"; "her significant other [someone in a close relationship other than marriage]" "Significantly [= suggesting greater or hidden information] he did not appear shocked when he heard the news". Scientists cannot use a word with so many meanings. However, they borrow it and restrict its sense to a very precise definition. Among scientists, "significant" has only one meaning: A difference is "significant" only if it is so large that, by statistically valid means, we would expect to see that difference fewer than one out of twenty times that we did the experiment. For instance, if we took twenty samples of 10 people each, half smokers and half not smokers, we might or might not detect a difference in lung cancer between each sample of smokers versus its comparable sample of non-smokers. You can easily appreciate that with very low numbers there is a high probability that the difference would be random. However, if we took a sample of 10,000 people and found a substantial difference, the likelihood that this difference would be random is very small, and the probability that we would find the same difference in a second or twentieth sample of 10,000 people correspondingly large. There are mathematical means of testing how small this likelihood is. Thus to a scientist, the word "significant" means ONLY, "This difference is so substantial that I would not expect to see it if I ran the test twenty times. Therefore there is something meaningfully different between the first group and the second group." Thus it becomes a quest to find the cause of the difference, the basis of designing a scientific experiment. Likewise, subatomic particles are described as having flavors, but not because anyone has ever tasted them. Quite simply, the particles have certain attributes to which the physicist

Richard Feynman wished to call attention. We define attributes in the known world through our senses, of which we have only five. He therefore used a term appropriate to one of the senses to indicate the relationship of the attribute to the particle. Thus scientists at work use language only in a highly restricted, specific sense, opening the door to confusion when in the lay world we allow greater variation and interpretation intuited by context.

Sometimes (often) scientific writing style, with its passive voice, past tense, and plural subjects, is very tedious: "It was observed that…" "We conclude that…" This is the unfortunate consequence of a discipline that as teachers we need to impose. Science students, like most teenagers and young adults, tend to focus on themselves. Science, however, depends on the work of our forerunners, is most often collaborative, and always depends on the investment of others. As a judge of high-school science fairs, I concluded that, while the winners (I mentored several and judged others) were clearly very capable young men and women who understood what they were doing and did it well, they often presented their work as their own, when it was not. I characterized their presentations as follows: "I worked on a project that no teenager had ever heard of, but someone told me that it was an important part of cancer. I used $5,000 worth of supplies, in a laboratory to which I had dubious legal access, and $200,000 worth of equipment, which had to be run by a graduate student because I was not permitted to tinker with it, and I found…" We also frequently have to remind them that they are obliged to list acknowledgements for their work.

When I introduced such students into my laboratory, I told them explicitly that they were guests, and that they would do the experiments that I directed them to do, not those of their own imagination. (We would of course discuss the reasons for the experiment and the options available.) To do otherwise would have been to cheat my sponsors. The university paid me to teach university students, and the taxpayers entrusted a government agency to distribute their money into research that the agency considered the most likely to benefit the country. There was no subsidy to humor high school students. If I were to allow such a student to follow his or her whim, I would have to devote time to learning enough about the project to mentor it, and I would be diverting resources purchased for other goals. Thus I would be cheating the university and the government.

We need to instill this sense of humility into beginning students. Often, however, it works too well, and first drafts of doctoral dissertations are often soporific. However, like the poet or artist who can feel free to innovate only after mastering the discipline of the classical style, we need students to learn that prizes are given to those who, at the end of the marathon, are barely a step ahead of a cohort of racers, all of whom have been building the enterprise. Furthermore, like a racecar driver or astronaut, the individual triumphs because a long history and army of supporters have led to the particular achievement. On our papers and publications we list the sponsors of the research. We need students to learn that. It is only the final lesson that we can teach maturing scientists, that they can lighten up on the style.

Even a mature scientist must approach his or her work with some humility, sense of humor, and understanding of public perception and expectation. In popular culture, a scientist is free to examine whatever strikes his or her fancy, to search for the "truth," whatever that is. In television crime dramas, the scientist is often heard extolling the primacy of "my science," usually in the context of human experimentation that otherwise would be illegal. This is not a fair representation. First of all, research today is sharply constrained by appropriate public concern as to what is moral or legal. We may sometimes dispute the judgment or bias of a public concern, and hopefully we can successfully defend our positions, but in principle there is no problem: all humans must cope with the expectations of human society. It was scientists who broached the theoretical (now largely considered to be overestimated, thanks to precautions recommended by the scientists) danger of the then nascent genetic engineering. (Genetic engineering is discussed in Chapter 9, p 114.) Second, our idle curiosity may lead us in many directions, but in reality science is expensive. Most science requires elaborate tools, supplies, computers, or expensive field trips. There are very few self-sufficient amateur (in the original sense) scientists; everyone else is subsidized, and we must win the subsidies. I might be very interested in the variation in the number of spots on ladybird beetles—and indeed I can think of a few valid reasons why this would be interesting—but, unless I can convince someone that this specific knowledge has value to humankind, my university or the government will not pay for it. Government agencies like the National Science Foundation and the National Institutes of Health appropriately set priorities as to what they consider to be of value, and scientists appropriately direct some attention toward seeking the funds that are offered. This is how government assures that its research investment targets government-desired goals.

What we ask is that the funding agencies set the questions to a large enough scope for us to make progress. A scientist's concern is that a governmental officer's (bureaucratic) expectation might be too narrow and thus miss the larger and more fundamental issue: "cure breast cancer," in other words identify the disease earlier and develop more effective surgical solutions, rather than identifying the biology of the cancer and stopping it before it starts, or killing it by using drugs specifically targeted to its unique biology. In scientists' parlance, to target poliomyelitis the funding agencies invested in research into the biology of viruses as well as in treatments then available and foreseeable. The elimination of polio was achieved by development of a vaccine, as opposed to the narrower but more immediate goal favored by some, which would have "developed the best iron lung in the world".

Even the development of a vaccine had a larger and a narrower goal. The larger goal explains the "coincidence" that three laboratories nearly simultaneously developed vaccines against polio. It was always possible to improve the iron lung since it was originally designed in 1928. The problem with developing a treatment for the disease itself was that it was nearly impossible to handle poliovirus in the laboratory. The virus infected only certain monkeys and humans. To study it, one would have to infect these monkeys and then attempt to do something before they died. Then, in 1949, John Enders, Thomas Weller, and Frederick Robbins demonstrated that they could grow cells from green monkeys in petri dishes and that they could infect these cells with poliovirus. Because they could do this, it then became possible to study the virus in the laboratory. Within three years, several laboratories learned how to tame the virus: by killing it gently (without deforming it) and using the dead virus for immunization (the Salk vaccine) or by mutating the virus so that it could not produce polio but could still cause the body to produce antibodies (the Sabin and Koprowski viruses). The public knows the names of Sabin and Salk (and Koprowski) (see Chapter 4, page 52) but Enders, Weller, and Robbins, rightly, were awarded a Nobel Prize.

In the USA, members of Congress frequently attack what appears to be useless or frivolous research. Senator William Proxmire (Senator from Wisconsin 1957-1989) used to produce "Golden Fleece Awards" for what he considered to be completely useless funded projects, and recent Presidential and Vice-Presidential candidates attacked wasting money on "fruit fly research in Paris, France" and DNA testing of grizzly bears. Sometimes scientists poorly describe and poorly defend their projects, but the projects, as judged by their peers in competitive access to grant funds, usually have validity. One year Proxmire criticized an inquiry into why small Islamic communities, when they came in close contact with western peoples, often became more fundamentalist (this was shortly before the Iranian revolution). That same year he criticized a project entitled "The Chemistry of Love". That project addressed the topic that people who habitually sought inappropriate or destructive partners had similarities to people with depressive psychiatric disease but, unlike the latter, did not respond to psychotherapeutic drugs. The writer of the application could have easily justified his research but did not effectively do so. Likewise, study of either Drosophila or, the apparent reference of the quoted phrase, a major pest of olive trees, has considerable value and has often yielded important returns; and identifying a limited number of bears and their relationships is a very efficient means of tracking and planning for the survival of an endangered species. But it is up to the scientists to make their argument clear.

One of the more bizarre accusations of recent times has been that issues such as human contribution to global warming are being argued "so that scientists can keep their research grants." Nobody likes to admit they are wrong, particularly if doing so has consequences—but this is a standard human flaw, open to everyone, including business people, doctors, lawyers...and politicians. Scientists are no more subject to this pressure than others. Grants subsidize research rather than scientists and, if the research leads to a dead end, the grant will most likely not be renewed. Beating a dead horse or barking up the wrong tree is not recommended as a means of competing for grant funds. Fraud, though a problem, is self-correcting: if the fraudulent claim is interesting or important enough to warrant future funding, others will try to replicate it, and claim will soon be challenged as a potential error of technique or interpretation.

Academic salaries do not approach those in potentially alternative professions, such as medicine, or alternative career paths, such as working in the biotechnical business. At best, a grant from the government replaces salary otherwise paid by the university or employs the scientist for research during the summer (typically 20-25% of salary). It would be hard to find a scientist, who, sanely, would argue that he or she went into the field for the money. While some of us might be (or are) fervent and passionate protagonists, by and large we are perfervid because we believe that our data support an important conclusion.

In sum, scientists do speak English (and other languages). However, our language is necessarily constrained for the very practical reason that we require precision beyond the tolerance of common speech. In the same sense, a machinist would not tolerate an apprentice who asked for a "thingamajigee" rather than a specific type of wrench, or a lawyer would challenge a witness who said "in the evening" rather than "about 8:15". The confusion arises when scientist and layman fail to recognize when language is restricted and when it is not, and neither clarifies the issue to the other. The most acrimonious of these disputes is the argument that the "theory of evolution" (more technically the theory of natural selection explaining the evolution of species) is "only a theory". To a scientist, a theory is not a guess but a hypothesis based on a very large basis of evidence, and very unlikely to be refuted. Like anything in science, any theory could be overturned by the unequivocal results of an experiment, but scientists, though typically not high-rolling gamblers, would be willing to place a lot of money on bets concerning the validity of a theory. But all of this has been discussed in the first book, *Born This Way: Becoming, Being, and Understanding Scientists* (available from several sources; see Endnote; hereafter referred to as BTW1)[7].

~~~~~

The life blood of the sciences: Hypotheses: Evidence, Logic, Falsification

To a scientist laws are observations so completely consistent that any apparently contradictory observation must be examined further for potential error or misinterpretation: the Law of Gravity and the Laws of Motion explain so many phenomena, from the movement of planets, comets, and satellites to the flow of winds and ocean currents and even some of the mechanics of chemical reactions, that we cannot deny the laws without creating a huge void of meaningful explanation. We use the laws in innumerable fashions every day, from designing roads, vehicles, and buildings, to making unconscious and uncalculated learned estimates of hitting or catching a baseball, approaching an intersection in a vehicle, balancing packages that we carry, and catching a wobbling vase before it falls. These are laws. Everything else is a theory or a hypothesis, a theory having the honor of greater (but undefined) higher certainty, and any experiment, if not refuted, can overturn it.

> "When the fact that one encounters opposes a reigning theory, one must accept the fact and abandon the theory, even that theory which, supported by great names, is generally accepted"—Claude Bernard.

> "In general we look for a new law by the following process. First we guess it. Then we compute the consequences of the guess to see what would be implied if this law that we guessed is right. Then we compare the result of the computation to nature, with experiment or experience, compare it directly with observation, to see if it works. If it disagrees with experiment it is wrong. In that simple statement is the key to science. It does not make any difference how beautiful your guess is. It does not make any difference how smart you are, who made the guess, or what his name is – if it disagrees with experiment it is wrong. That is all there is to it."—Richard Feynman, **The Character of Physical Law (1965).**

> "The exception proves that the rule is wrong." That is the principle of science. If there is an exception to any rule, and if it can be proved by observation, that rule is wrong.—Richard Feynman

> "No amount of experimentation can ever prove me right; a single experiment can prove me wrong."—Albert Einstein.

Before we get into that, however, let's consider the basic structure, the hypothesis.

Everyone makes hypotheses and often tests them. Therefore everyone is, or potentially is, a scientist, and scientific reasoning is a perfectly human and comfortable activity, not something arcane, weird, or geeky. We can look at several examples. The lamp is out. We will start with a basic theory, one in which we have confidence because we and others have tested it many times and never found it to be wrong, that electricity must flow in a continuous circuit, that is from one of the wires that comes into our house (the "hot" wire) through the circuit breaker into the slit that takes one prong of our plug, through the wire that goes to the lamp (bulb), the switch that controls the lamp, into the contact at the bottom of the lamp, through the wire in the lamp, out through the metal screw part of the lamp, back down the wire, into the other prong of the plug, and so forth back out of the house. We will make a hypothesis, based on the theory, that the circuit is interrupted. (Technically, since we hypothesize mechanisms that explain phenomena, we hypothesize that a complete circuit carries sufficient electricity to light a bulb, and that the failure of the bulb to light indicates a break in the circuit. Obviously this hypothesis subsumes many others, such as what electricity is and how it can cause a bulb to glow, but that is a more complex issue.) Based on other evidence in our experience, for instance that light bulbs fail far more frequently than wires, switches, or circuit breakers, we will conduct an experiment: replace the light bulb. (We could also look for further evidence, for instance looking to see if the filament is broken.) If the lamp now works, our hypothesis is verified and the theory holds. If the lamp does not work, our hypothesis is falsified and we move within the constraints of the theory to consider other possible breaks, in the switch, wire, outlet, circuit breaker, and even the possibility of a general blackout. These are alternative hypotheses. Only if we can demonstrate that there are no breaks anywhere in the circuit can we consider that the theory is wrong. Note that this rather trivial and obvious example is very limited as a hypothesis. Specifically, it does not truly postulate a new general mechanism to explain the relationship between two classes of events (e.g., the flipping of switches and the lighting of bulbs), but it illustrates the point.

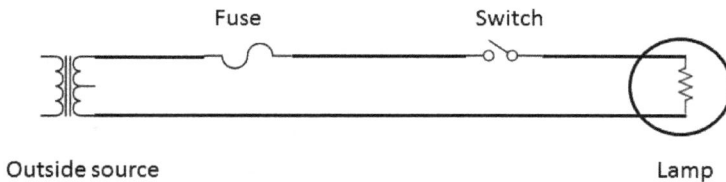

Figure 1.4. An electrical circuit. The current, consisting of moving electrons, has to complete a circuit. Any break in the circuit--a blown fuse, an open switch, or a broken filament in the lamp, will interrupt the flow of current, and the lamp will not glow.

Note that there is nothing structurally that differentiates this basic chore from a scientific experiment, and that therefore we are all scientists. It differs only in that it does not extend the theory or expand our knowledge; it merely resolves a

problem. Being a scientist means that one wants to extend theories into new territories, to explain further phenomena and even to overturn the theory should it prove inadequate. Another example: A ship inside a narrow-necked bottle. Assuming that there is something equivalent to a law about the ship passing through the side of the bottle, we can generate many hypotheses: That the base of the bottle was put on after the ship was inserted; that the neck of the bottle was heated and drawn out after the ship was placed in a large tube of glass; that the ship was painstakingly assembled using chopstick-like forceps through the neck of the bottle; that the ship is collapsible, was assembled and folded into a very compact form, inserted into the bottle, and then erected once inside; and perhaps other hypotheses. We would then marshal evidence, by looking for seams in the glass, moveable parts on the boat, *etc.* to try to prove one hypothesis or another.

Or rather, disprove one hypothesis or another. It might seem counter-intuitive, but it is not possible to prove a hypothesis, only to disprove all alternatives and thus leave one standing. For instance, we could make many hypotheses as to why the northern hemisphere gets warmer towards June and July: the longer day length gives the sun more time to heat the land; the higher position of the sun in the sky provides more direct heating; the warmth and the day length both run in cycles but are independent of each other; the names of the months control the temperatures, with months containing an R being colder; green leaves absorb or generate heat; or even that the spontaneous warming and cooling of the earth changes the length of the day. By logic or experiment, we can eliminate most of these hypotheses. But even if the first two remain standing, we can never prove that there is not a third factor, even God, that causes these to occur coincidentally together even though there is no causal connection. This might seem a little confusing, but it is the same problem that I would have in arguing that, since I ate a piece of chocolate pie yesterday and got sick today, the pie caused me to get sick. Coincidence is not proof, an issue frequently argued in popular medicine. For instance, recently the argument was made that, since autism is diagnosed after children are vaccinated, the vaccination caused autism. Considerable effort was made to disprove this hypothesis[8] (autism, a behavioral disorder, is not recognizable until children should begin to show certain types of interaction with others, an age that follows the age at which vaccinations are customarily given), but we cannot rule out the possibility that another factor—bubble gum?—will not eventually prove to be a causal agent. Thus we strive to <u>eliminate competing hypotheses</u>, leaving one standing by virtue of logic and evidence, and can never be fully certain that it is valid.

Concerning the ship in the bottle, we can take as a scientific law the observation that matter cannot pass through other matter. This is not strictly true on an atomic level, but it will serve our purpose. Scientific hypotheses, theories, and

laws differ only in the level of certainty that we ascribe to them. Hypotheses are conjectures concerning the mechanism of action, based on our logic and the information (evidence) that we have in hand. We hypothesize that a rather slight, older man was killed by a big, burly man who possessed a knife and had a grudge, but we wait for further proof as presented at trial: the legal equivalent of a scientific experiment. The experiment would differ only in the sense that it would seek a generalized hypothesis, that it is likely that big, burly men with knives and grudges will be able to kill weaker men. (This isn't so ridiculous if we change the circumstances, for instance that, when male bison compete for dominance of the herd by slamming into each other, the larger, heavier bison will almost always win.) A hypothesis becomes a theory when it has not only been verified in an appropriate experiment, but it has been verified many times over in many circumstances, such that we would be genuinely surprised by a failure to verify it. In 2011, a claim was made that a neutron had travelled faster than the speed of light. Such a finding would counter Einstein's Special Theory of Relativity. The scientists who made the claim pleaded with other scientists to find what, if anything, was wrong with their finding, and the discovery made the front pages of newspapers around the world. Sure enough, it now appears that a problem with the circuitry led to an inaccurate reading, and that Einstein's theory remains intact (BTW1,[7]). A theory becomes a law when we really have no hope of ever seeing anything that contradicts it. For instance, buildings do not levitate (Law of Gravity) and objects at rest do not begin to move unless there is something that moves them (Newton's First Law of Motion).

~~~~~

A hypothesis is a proposition of a mechanism that can explain the relationship between two phenomena. It is a valid hypothesis only if it can be tested, that is to say that there is a means of proving it false. This somewhat bizarre rephrasing is necessary, because a false hypothesis of mechanism can still lead to the predicted results. "If the warming of the earth causes the day length to get longer, then the warmer the earth, the longer will be the day length" will in general predict longer days in summer and shorter days in winter. To test a hypothesis we must therefore try to disprove it, or falsify it, observing its validity by an "if or only if" criterion.

The "if and only if" rule applies to scientific hypotheses. To speak as a scientist, it would be called "falsification," a term used because technically it is not possible to prove a hypothesis, only to disprove it. This is a source of huge confusion between scientists and lay people, as exemplified by a jury scenario: "Dr. Jones, is it possible that this DNA sample came from someone other than my client?" "Well, the chance is less than one in one billion, and there are only 350 million people in the country." "I didn't ask you that. Is it possible?" "Well, yes, it is possible." "Thank you, Dr. Jones, that will be all."

Why is it not possible to prove an argument? Let's look at an example. Snow melts on warm days. I could make two hypotheses concerning the mechanism: that the warmth melts the snow, or that melting snow warms the air. From the correlation between heat and snow melting, either hypothesis would be valid. The lack of a plausible mechanism for the latter would be an argument against it, but that is another aspect of understanding hypotheses. We can devise an experiment in which we could separate air temperature from snow melt, for instance by putting the snow on a warm plate while simultaneously blowing cold air over it. We could arrange such an experiment to show that melting snow does not warm the air. (In fact snow melts by absorbing heat, which will cool the air.) Thus we could show that melting snow will not warm the air, and thus rule out that hypothesis. However, ruling out one hypothesis does not absolutely prove the point. It could be that another factor, which by chance or mechanism always accompanies warm air, is the actual cause of snow melting. For instance, a volatile solvent in the air could penetrate the snow and lower the temperature above which it melts, thus causing it to melt. The pressure of the air on top of the snow can affect the temperature at which it melts. That's how ice skates work: the weight of the body concentrated into the narrow surface of the blade puts considerable pressure on the ice underneath the blade, causing the ice to melt and us to glide smoothly along. It is also how glaciers slip down mountainsides. The point is this: We can rule out some hypotheses—we can falsify them—but we cannot prove that, somewhere, there is a new hypothesis (radioactivity heating the earth?) that will also and perhaps better explain the correlation. We can only work to a point at which no one can think of any other explanation, and numerous tests of the hypothesis have failed to disprove it. At that point, pretty much undefined but sensed, we can describe the hypothesis as a theory, for instance the theory of natural selection. We are still working out the details of the mechanisms and, like any theory, it is always possible that something will come along to make us question it. However, it explains so much, there is so much evidence for it, and denying it would unravel so many other hypotheses, that no scientist truly expects the theory to collapse[7] (BTW1). If there has never been any physical evidence to the contrary, and the theory seems to work in all circumstances, at a microscopic as well as planetary level, and we would be truly amazed to see a contradiction, such as a tree or car without obvious interference suddenly lift into space, then the theory becomes a law or first principle, such as the Law of Gravity.

There are many examples of the "if and only if" principle. Two of the most famous include Pasteur's argument against the spontaneous generation of life (p. 23)— itself an extension of Francesco Redi's documentation[9] that maggots did not generate spontaneously in rotting meat—and Joseph Goldberger's demonstration that the disease pellagra was caused by dietary deficiency (p. 29), later determined as vitamin deficiency, rather than infection of genetic flaw. These are the subjects of the next chapter.

STATISTICS

In most situations involving humans, we cannot create an experiment in which we can effectively oppose cause and effect. Even with our best animal models, it is very rare that a given treatment will produce cancer in 100% of the exposed animals, or that a given treatment will destroy the cancer 100% of the time. With animal models, we try to work to a point at which the difference between a control and an experiment is obvious, but it often is not as obvious as we would hope; and in the case of humans, it is often not at all immediately obvious. Therefore we have to postulate cause and effect based on a statistical evaluation, that is, a calculation that says that, if we repeated the same experiment 100 times, we would get the result five or fewer times. This does not prove causality. It merely says that the correlation is unexpected and may have been created by the putative cause.

Generally, statistical probability is calculated on the presumption that values distribute according to a Poisson Distribution, or bell curve (Fig. 1.5). The bell curve describes a trait that can be measured but which varies continuously in a population, such as weight or height. Many such continuously varying parameters are distributed in this fashion: Most individuals are found near the mean of the variable, and the greater distance from the mean, the fewer individuals are found. In the US, the more the measurement of male height is above or below 5' 10" or female height is above or below 5' 4", the fewer individuals will be found at that height.

The Poisson Distribution does not describe variables that are essentially discrete or discontinuous. For instance, there is no continuous distribution along a gradient between "female-like" and "male-like" (Fig. 1.5 right). Almost all individuals are clustered in one or two categories. On a more complex level, because of the historically late meeting of various groups of humans, in some societies skin color might be discontinuous while in other societies, such as in Brazil or Hawaii, it might tend towards a continuous distribution. It is important to realize that bell curves do not describe all situations; sometimes distributions are skewed for one reason or another, and that the simplest statistical assumptions such as these are not necessarily valid.

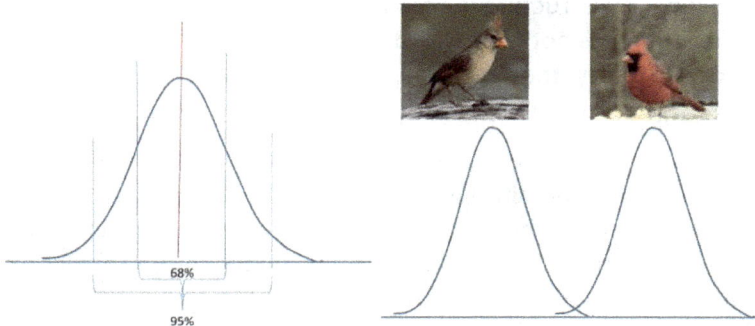

Figure 1.5. Left: A Poisson distribution or bell curve. The ordinate, or vertical axis, represents the number of individuals found at any given quantitative attribute, such as height, represented on the horizontal axis, or abscissa. For any randomly distributed quantitative attribute, most of the examples will fall in the middle of the range, with relatively fewer individuals encountered the farther one deviates from the middle of the range. Such patterns are sufficiently consistent that one can calculate the range in which approximately two-thirds of the individuals will be found (defined as one standard deviation from the mean) and the range in which 95% of the individuals will be found (two standard deviations). The numbers come from properties of natural logarithms. Scientists define a point as "statistically significant" if it falls outside of the 95% range. By this they mean that it is worth examining further to see if there is a cause, but by definition 5% of any population will be found above or below the 95% range. Right: not all measurements can be forced into a single Poisson distribution. As is true of many sex characteristics, "female" (left) and "male" right colorations of cardinals distribute into two means, with very few individuals displaying truly intermediate coloration.

This is the most mathematical of restrictions that we encounter. Others are more fundamental, and they relate to the most common misuses of scientific reasoning, testimonials and false associations of logic. You can find examples of these nearly every day.

TESTIMONIALS

The testimonial is the single example, most commonly in first person: "I followed Dr. J's diet plan, and I lost 50 pounds!" It is important to remember that the function of such a testimonial is to give a single example to entice the listener to generalize from that example to a general rule, from which the listener can deduce the effect on himself or herself: Testimonial-giver X lost 50 pounds. Therefore all people who follow Dr. J's plan lose 50 lbs. Therefore if I follow Dr. J's plan, I will lose 50 pounds.

This is the logic of a false syllogism. A true syllogism allows one to deduce an individual truth from a general truth: If all girls are pretty, and if all children named Mary are girls, then Mary is pretty. A syllogism, basically an issue of logic, works only in this fashion, and reaches only one conclusion. It does not work backwards: If a flower is pretty, it does not follow that a flower is a girl. In other words, the logical flow does not of itself establish the "if and only if" structure of a true scientific experiment.

First, the assumption may be false. The statement "if all girls are pretty" may be false in fact even if the structure is acceptable. If the answer to "if all girls are pretty" is "no," then it is not necessary that Mary be pretty. She might be, but if some girls are not pretty, then Mary might be one of those girls.

Second, the problem with a testimonial is that it attempts to impute the syllogism where none exists, or causality from coincidence. In large populations, many things are possible, with or without cause. Someone will win the lottery, but the fact that this person chose the numerical version of his son's birthday as numbers does not create the prediction that choosing birthdays will win again. Likewise, a certain number of people will be involved in traffic accidents. Many of these will have proximate causes, but the fact that you were driving on the priority road when someone pulled out from a stop sign without seeing you does not necessarily make you a less worthy driver. You may or may not have been less attentive than you should have been. Likewise, the dieter giving the testimonial may have been enormously motivated by any of several social or financial consequences of his obesity, or a serious medical problem. In this circumstance almost any effort to lose weight would probably have worked. The causal relationship between Dr. J's diet and his weight loss is not proven, and it will not necessarily work for you or anyone else. Other examples include non-prescription medicines for conditions that spontaneously resolve themselves. The old joke is that this treatment will cure a cold in seven days. Without treatment, the cold can last a week!

The only valid predictor is the statistical measurement, done with suitable controls. If 1000 people follow Dr. J's diet and are compared to 1000 people who make no effort to lose weight and 1000 people who try another procedure (exercise or another diet) and, for instance, 200 of those on Dr. J's diet lose 10 or more pounds, compared to 20 who lose weight while doing nothing and 30 who lose weight doing something else, then one can conclude that the diet has some benefit. Note, however, that there is still a catch: 800 people who followed Dr. J's diet nevertheless did not lose weight. It will not work for everyone. In more dismal terms, if 90% of patients survive a difficult operation, the loss of 10% does not necessarily indicate failure or malpractice on the part of the surgeon.

Thus the testimonial attempts to impute the relationship between one instance and another instance. It does not establish any greater causality than my winning the lottery by choosing numbers based on the license plate of a car stopped ahead of me at a red light.

There is the other extremely difficult problem when dealing with large populations: since we cannot truly control the situation, there are huge numbers of variables that potentially can affect results. Suppose we tried to compare rates of heart disease among Mexican-Americans in Los Angeles and among people of Swedish descent in Minnesota. Using your hands and feet, you can readily count

off the numerous likely differences in genetic background, diet, probability of smoking, exposure to sunlight, exposure to childhood diseases, exposure to airborne carcinogens, amount of exercise, level of education, probability that they have lived in the same location all their lives—and we have not begun to consider the age and sex of the subjects. If there is any social component, for instance if the data are gathered by interview, there may be other differences. It is well known that different ethnic groups give different responses depending on circumstances such as the race of the interviewer. Even if one relies on more solid data, such as hospital admissions, some groups are more likely than others to present themselves to formal medicine at an early stage of a disease. If one attempts to conduct a survey in front of a sports bar at night; in front of a school in a prosperous neighborhood when parents are dropping off children; in a supermarket at different times during the day; in front of a factory; with workers exiting from a high-end office building; with customers at a "dollar" store; etc., it is likely that the results will vary. Correlation is NOT causality, and statistical analyses have to be very carefully planned.

In the ideal laboratory setting it is possible to so tightly limit the variables and to propose so many variants that we can come close to the ideal "if and only if" criterion that approaches certainty, so that we can publish papers in which we argue that A causes B—the basis of a hypothesis of mechanism that works toward being a theory. However, in many laboratory situations and commonly in the real world, we cannot establish that level of control. For instance, to predict the outcome of an election, we might wish to interview 100 people about their intentions. However, if as described in the paragraph above we choose to do our interviews in front of a school in a prosperous neighborhood where parents are dropping off their children; of customers coming to buy the latest gadget at a high tech store; at a baseball game; at a hockey game; on the subway; in front of a supermarket at 8 AM, 2 PM, or 7 PM; in front of the courthouse—well, obviously, we are going to be interviewing different types of people and are likely to get very different results.

We often have to work using statistics, but these statistics will be sharply, even severely, constrained by the requirement that we achieve the closest approximation to randomness that we can achieve. Two factors are major issues. One is sample size, and the other is randomness. After all, we are going to try to use correlation to argue causality. The issue of sample size is fairly obvious though actually addressed by mathematical rules. If we flip a coin, we should have a 50-

50 chance of getting "heads". If out of six flips, we get 4:2, 5:1, or even 6:0, we will not be terribly perturbed, but if we flipped the coin 6,000 times and got 5,000:1,000 we would suspect that the coin was perversely weighted or that some other factor was biasing the results. Suffice it to say that there are mathematical means of calculating the probability that we would get such a result by chance alone, and if that probability is less than 5% (if we replicated the 6,000 tosses one hundred times and would expect that such a result would occur fewer than 5 of the 100 times) then we consider the result to be "statistically significant" or worthy of investigation as to whether there is something biasing the result. Note again that "significant" does not mean "portentous" (as in, "the falling star was read as a significant omen"), "capable of having an effect" (as in "there is no significant amount of salt in this meal"), or "important" (as in "significant other"). To a scientist, "significant" means only "unlikely to occur more than 5% of the time in many replications of the experiment". Randomness is much more difficult to achieve, and the problem is not limited to social surveys. For instance, most studies indicate that smokers are more likely to develop lung cancer than non-smokers. Does this prove that smoking causes lung cancer? No. Originally, most of the smokers were men. Smokers may have very different life styles from non-smokers. Let's postulate that they may be more nervous, or occupied in more stress-inducing jobs, than non- smokers. Smoking may be more an urban than rural phenomenon and therefore smokers also are exposed to more pollutants in the air. Smokers may on average be more or less obese than non-smokers. Typically, smokers drink more alcohol than non-smokers. Non-smokers may belong to certain religious or social groups that impose many other restrictions on their lifestyle, any of which may be a factor. The mean age of smokers may differ from that of non-smokers. Since smoking has changed with time, different generations may have been exposed to very different experiences in childhood. The closest that we can come to certainty is to have (a) large numbers of studies (b) involving large numbers of individuals (c) as closely matched as possible, according to age, weight, sex, race, economic factors, and any other criterion that we can identify. Finally, in the laboratory, concerning the mechanism, we can with highly inbred mice demonstrate that some of the constituents of tobacco smoke produce the same types of lesions, including cancer, in the lungs of mice that they we see in humans, and we can demonstrate that these chemicals bind to DNA and damage it in a way that can cause cancer. Even then, we cannot say that we have proved that smoking causes cancer. What we can say is that, if you are gambling, the odds for smoking to have something to do with lung cancer are extremely good. Thus scientists object to news reports by saying that the numbers are too small (the fact that 20 patients developed cancer when only 18 were expected could be within the range of natural variation) or that the experiments were poorly controlled (people who eat a large amount of quinoa—a nutritious grain-like seed from Peru—very likely have many other differences in diet and habit

from those who have never heard of it). And finally, scientists object vociferously to "anecdotal" evidence. The coincidence of the two incidents does not establish causality. To carry the point further, cancer takes many years to develop even in a heavy smoker. Suppose that over these many years our smoker has several times changed his preferred brand, and has many times "mooched" cigarettes from smokers of many other brands. Should he now develop cancer, is he in a position to say, "The Gaulois cigarette that I smoked when I was a soldier after WWII in France caused my cancer"? Of course not. There is no way to assign the mutation that led to the cancer to any single cigarette that this gentleman smoked in his lifetime. And yet this type of argument appears, with more subtlety, quite often. One sees it in the newspapers and on television, in the guise of "this airport opened five years ago, and now I and my neighbor have cancer". To establish any semblance of causality, one would need to have adequate statistical data by all the criteria listed above, and a solid argument as to the mechanism. Otherwise the incident is an "anecdote," one case of cancer of the estimated 1.6 million new cases per year in the US, that happened to occur near an airport.

At one point I consulted on the closure of a toxic landfill. The task involved rearranging the shape of the mound (now a large hill) of trash in the landfill so that water would run off of it without accumulating, putting a large rubberized cap over it all, and then adding new dirt so that the surface would at least be grassland, suitable for a light-use park. The people in the neighborhood were of course concerned about the health risks involved in shifting materials around in the landfill, and requested health studies. They wanted testing for cancer. I and others had the obligation of explaining to them that cancers take years to develop, so that the evaluations would come too slowly. Furthermore, we would have to evaluate the incidence of cancer against some sort of control, such as a comparable neighborhood. However, this neighborhood had a high rate of turnover, was also near to a major airport and highway, both of which could produce carcinogenic fumes, and had a very mixed racial and ethnic population. To whom could we compare these people? We finally agreed that the most useful thing to do would be to install air quality monitors, looking for known toxins, with arrangements to shut down the operation if anything serious was detected, and further to monitor admissions for asthma at local hospitals and prescriptions for asthma medications at local pharmacies. (Asthma of course has nothing to do with cancer, but it is a quick-response indicator of poor air quality. Unlike a random or unfocused ("shotgun") screen for cancer, a screen for asthma had a real possibility of protecting the population in real time.)

IMPLICATIONS OF CONFUSING CORRELATION AND CAUSALITY

The reason that this discussion is important to us is that it is precisely this type of misuse of data—anecdotal arguments or confusion of correlation with causality—that is so often used to argue against the theory of evolution or, at a more destructive level, to justify a particularly heinous or cruel attitude toward fellow human beings. One can readily argue that there is a direct line from Darwin to Hitler, and that most of the cruelest political activities of the 20th C were based on somewhat innocent to intentional distortions of the meaning of evolutionary theory.[7]

BIAS IN SCIENCE

It is easy to blame society and leaders for misinterpreting and misusing the findings of a politically neutral science, but the explanation is not that easy. First, science is never totally politically neutral. As certain facts make no sense until we have a logic to explain them, scientists are humans and operate within the assumptions of the society. Thus, as is described below (p.37), primatologists based on social biases overrated the importance of the alpha male and missed the importance of female apes in establishing their own social hierarchies. In many cases dealing with human evolution and selection, although ultimately science proved the interpretations wrong, well-meaning, intelligent scientists sometimes led the way in arguing for a hierarchical ranking of human races. Even well-meaning scientists sometimes still drift into these traps. In 2002, a popular textbook illustrated a chart of human evolution that suggested that all human variants derived from an ancestral African type. This is a valid statement. However, it does not follow that all Africans resemble that ancestor, as the graph implied. Most likely, the ancestral humans more closely resembled today's Khoisan peoples, and humans within Africa diverged as much or more from the

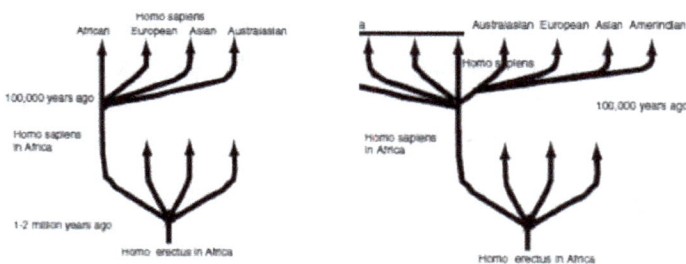

ancestral type as did those humans who left Africa and continued to evolve.

Figure 1.6. Unintended bias readily creeps into our processing of data. The lower part of the figure on the left suggests that, of the several variants of *H. erectus* that arose in Africa, only one lineage gave rise to modern humans (reasonable enough). However, the upper part suggests that, of this lineage, modern humans developed, with the African group closest to the original progenitor stock, with the European, Asian, and Australasian (aborigine) peoples more and more deviant from that progenitor stock--thus implicitly suggesting a hierarchy of human races. The reality is more likely as represented on the right: many branches within Africa, which harbors the greatest diversity of human genes, with Australasians probably closer to one of the African branches, and Asians deriving from the European branches. An even more accurate scientific version would use the lengths of the branches to indicate the relative affinities of the groups.

It is the responsibility of scientists to assess the social implications of what they do and say, and to consider the possibility that expertise in the laboratory is not expertise in social policy. Similarly, it is the responsibility of those who are not doing the science to remember the social imperatives and to be sufficiently aware of the fallibility of science to resist making or asking for bad law in the name of science.[10]

~~~~~

Chapter 2: The tools of the trade: Experimental design

The heart of science is the experiment. Many brilliant experiments are being done in many laboratories every day, and a mark of very high praise in a meeting or in conversation is, "That's a really nice experiment!" (often "a cool experiment" in the highest sense of popular use of the word, and expressing real admiration for the cleverness with which the experiment resolves an issue). Such an experiment in one swoop definitively rules out a plausible alternative or creates a situation so otherwise unpredictable that it becomes extremely difficult to argue against either the "if" or "only if" portion. The best way to picture such an experiment is through example. Some of these experiments, such as those that gave rise to molecular biology, can be described in such a way that the brilliance of their design, logic, and execution is palpable, and we will look at these in a later chapter; but others rely on abstruse information, techniques, or hypotheses, or require a mountain of preliminary explanation to become accessible. It is better to start with classical experiments, done before technical complexity and required background became so daunting that the field could be described as hermetic.

Documenting that life does not spontaneously arise

A few years before van Leeuwenhoek built his first microscope, it was not at all obvious or even reasonable that there existed life that we cannot see. This ignorance extended into the realm that we today would not necessarily consider to be microscopic. The human eye can resolve two dots to a limit of approximately 0.2 mm or 3/64 of an inch; below that limit, the dots blur into one. A fly egg is approximately 0.1 mm in diameter and can be easily seen only when its white color contrasts against a relatively homogeneous background such as the medium in which laboratory-raised flies lay their eggs. They are not easily detected against more confused backgrounds. Thus it was no wonder that most people thought that maggots spontaneously arose on meat that was exposed to air, as frogs appeared in ponds after rains and small worms appeared in standing water. However, in 1651 William Harvey, who first recognized the circulation of the blood, extrapolated from what he could see to a general rule in which he argued that "Everything comes from eggs," meaning that, as one could find an egg or seed stage for many or most animals and plants, he surmised that eggs would be found for all other kinds of creatures. Francesco Redi, noting that the maggots eventually turned into flies and that flies were common on meat, wondered whether there was a connection and in 1668 published a book entitled "Experiments on the Generation of Insects". In it, he outlined a careful series of experiments that convinced him that the maggots were not spontaneously generated but arose from the flies that circled around the meat (his hypothesis— that maggots arose from flies rather than spontaneous generation):

"I put in four flasks with wide mouths one sneak [snake], some fish of river, four small eels of Arno river and a piece of calf and I locked very well the mouths of the flasks with paper and string. Afterward I placed in other four flasks the same things and left the mouths of flasks open. Short time later the meat and the fishes inside the open flasks became verminous, and after three weeks I saw many flies around these flasks, but in the locked ones I never seen a worm". [11]

This experiment could be challenged. The permeability of the paper to air was not known, and it could be argued that the air in the flasks would "go stale" and not be able to support the spontaneous generation of maggots. He therefore modified this experiment and covered the flasks with a fine cloth, through which air could obviously pass. He used different types of meat, and tried the experiment at different times and temperatures. Finally, after demonstrating that meat could go approximately two weeks without generating maggots, he removed the covers and showed that, when flies could reach the meat, it quickly generated maggots.

His overall experiments contained several elements of good experimental design. He used several kinds of meat and several conditions. Thus his conclusions were not limited to "maggots do not generate on a dead snake when it is covered with cloth and left outside on a cool day in July"; he could generalize past specific local conditions and types of meat, allowing him to hypothesize that the rule was general: maggots do not spontaneously generate on decaying matter if flies are prevented from landing there. He also, over time, repeated the experiment several times, confirming that the results could not be attributed to a specific quirk of the experiment. For instance, flies might not fly on a very cold day or might not be able to land on a very windy day. Finally, he conducted a control experiment, one in which all conditions were identical except for one: the ability of the flies to land on the meat.

The control experiment is extremely important and is difficult to do correctly. For instance, if one injects a drug into a mouse and measures a specific effect, one must also inject the solution in which the drug was dissolved (in this instance without the drug) into a second mouse. Injecting something into a mouse frightens it and causes it pain, which can cause its stress hormones to change and affect its behavior, growth, ovulatory pattern, or other factors. Even the fright that other mice feel when they hear the squeals of the mouse that is the subject of the experiment can alter their behavior and the results of some experiments. All sorts of things, if uncontrolled or unmeasured, may prove to have affected a given experiment. A seasonal variation in fat metabolism, observed in the 1970's in mice that had been raised in the laboratory since the 1920's—hence over 300 generations away from the influence of seasons—was traced to seasonal variation in the nutritional value of grain used to prepare standard, commercially-purchased, laboratory chow for the mice.

Bacteria were another matter. They were not truly identified until the 19th C, and no one seriously thought that such tiny creatures could cause disease. By the mid-19th C, Pasteur had established that spoilage of food and disease among the grapevines of France, both issues of great practical and commercial importance, could be associated with microorganisms. The question then arose as to whether it was possible to prevent these microorganisms from attacking food and vines. If the infections were communicable, meaning that they moved from an infected organism or source to another, then it might be possible to prevent their spread. If, on the other hand, bacteria arose spontaneously (spontaneous generation) then the problem became much more complex. It was a question of such importance that the French Academy of Sciences offered a prize, the Alhumbert Prize, for the best proof of whether or not putrefaction could generate spontaneously or required an agent.

Pasteur had demonstrated that boiling milk would delay its putrefaction, but milk is obviously changed by boiling, and one could argue that it was so altered that it could no longer support putrefaction. (Pasteurization, originally used to destroy tuberculosis-causing organisms, consists of heating milk to high but not boiling temperatures, allowing it to cool, letting it sit to allow heat-resistant spores of bacteria (primarily at the time tuberculosis) to germinate, and reheating the milk to destroy this second generation of bacteria. The milk is less obviously cooked or altered, but it is far from sterile and will eventually spoil.) He sealed the containers after boiling and showed that the milk could be kept, but one could argue that the air was also damaged by the boiling and could not support generation of bacteria. Thus the contest was announced. Pasteur set out to prove that bacteria could not generate spontaneously, while Félix Archimède Pouchet attempted to prove that they could. The hypotheses were as follows:

> 1. Bacteria arose spontaneously but required undamaged (uncooked) food to grow.

> 2. Bacteria arose spontaneously but required something from the air to grow, and whatever was in the air could be destroyed by heat.

> 3. Bacteria arose from other bacteria that could easily contaminate even a clean preparation.

Pasteur considered also how bacteria might travel. Everyone has seen little bits of dust, illuminated by a beam of sunlight, floating in the air, and Pasteur considered that bacteria might move in the same fashion. Lazzaro Spallanzani had demonstrated almost 100 years previously that small animals could not generate in flasks that had been subjected to boiling and sealed, but that they could once the seal had been broken. The argument was that the air had been damaged and could not support spontaneous generation. Therefore the issue was to prevent spontaneous generation in a medium exposed to undamaged air.

Hypothesizing that the bacteria were carried by air, Pasteur endeavored to create a situation in which the medium (broth) was exposed to undamaged air, but that the air was so still that the bacteria would not be wafted into it. He finally designed the flask shown in Figure 2.1. He put broth, a sort of beef bouillon, into the flask, boiled it to kill the bacteria, and drew out the neck of the flask into a long S shape. He hypothesized that bacteria falling into the open end of the neck would settle into the lower curve and that the air would be too still to carry the bacteria up the neck and into the flask. It was already known, from experiments with vacuums and smelly gases like ammonia, that air itself could diffuse without a breeze, but that dust would settle out. He also put some of the boiled bouillon into an open flask, where it quickly became contaminated, demonstrating that the bouillon itself was not damaged by the boiling.

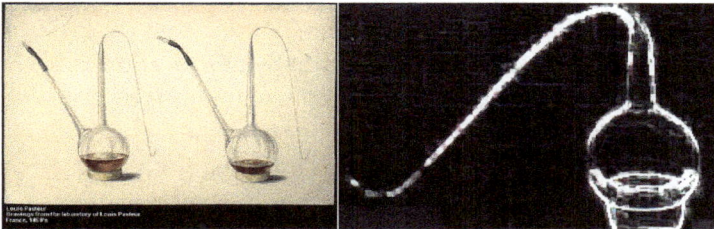

Figure 2.1 Left: Illustrations of Pasteur's flasks as indicated in his notes (the left tube was used to inject material into the flask and was subsequently closed). Right: One of his flasks still found at the Institut Pasteur, and still uncontaminated.

His flask did not become contaminated. To prove that there was nothing further wrong with his flask other than the ability of the bacteria to reach the broth, after a suitable period of time he tipped the flask so that the broth could reach the final curvature, and then returned it to its original position. Since the bouillon had now sloshed some of the bacteria back into the flask, it quickly became contaminated and putrefied. Other flasks that were not tipped remained sterile; indeed, one remains on exhibit, 150 years later, and still uncontaminated. Pasteur was awarded the prize, and today thousands of laboratories use petri dishes, based on the same principle, to study microorganisms in uncontaminated conditions (Fig. 2.2).

Figure 2.2. A petri dish in its proper storage configuration. Bacteria grow on the agar medium (yellow in smaller, upper component) which is inverted into the lid on the bottom. In this fashion the petri dish replicates the Pasteur flask, in that any contamination that might fall on the lid will not be carried upward onto the agar.

The "If and Only If" logic of the experiment is fairly straightforward. It is also of interest to note that, although Pasteur was right and won the prize, the specific conditions in which he did his experiment were crucial and lucky. If he had used an extract of hay (hay infusoria) as Pouchet did, his boiling would have failed to kill all the spores, and the broth would have eventually become contaminated. We know today that to kill spores in hay one needs much more drastic boiling.

A similar instance of the "If and Only If" logic is Sir John Snow's demonstration that cholera was a disease caused by an organism that lived in water.

Sir John Snow and Cholera

This classic study technically was not a full experiment but it used the logic of science to demonstrate that cholera was an infectious disease. Cholera was a devastating disease. Essentially a severe diarrhea, but one that could drain so much fluid from a person that it could kill a person by dehydration in a few hours, it would break out in cities and spread rapidly, killing hundreds or even thousands in the space of a few weeks. There were two major hypotheses as to what caused it: "Effluvia," by which was meant odors or gases escaping from infected patients, who thus poisoned the air for healthy individuals; or biological or chemical agents in the bodies of the victims. Snow therefore looked at the logic of what the evidence was telling him:

- Cholera traveled from city to city at the same rate that people traveled. Thus, if cholera broke out in Rome or Paris, it would not reach London faster than the time that it took stagecoaches or boats to reach London.

- If cholera came from another country, it would be seen first at a seaport. It would not appear suddenly in the Midlands of England.

- It could break out on ships, but only if the ships came from cholera-infected countries. If cholera had broken out in Rome, a ship coming from Rome might develop cholera, but cholera would not appear on a ship coming from Stockholm.

All of this evidence suggested that cholera was transmitted from person to person, but it still did not resolve the two hypotheses. But then Snow encountered a new patient who had no personal contact with any cholera victim. However Snow, an astute observer, learned that the patient had received clothes from a recent cholera victim. This was not unusual; if someone died young, the clothes were often recycled. Snow then refined his hypotheses:

- Hypothesis: If cholera is passed by effluvia, then all persons in contact with patient should get cholera, and those in contact with only the clothes should not.

- Hypothesis: If cholera is passed by liquids, then those in contact with the liquids should get cholera, whether or not the patient is present.

Since in a medical situation one does not usually have recourse to a lab experiment, Snow re-examined his evidence to see if the evidence supported one or the other of these hypotheses.

- It might be disgusting, but it was an issue that physicians could note. In the later stages of cholera, patients have vomited and lost through diarrhea so much that their digestive tracts are empty, and anything further that is lost is clear and watery, and may not even be noticed. In such conditions, Snow considered the possibility that the last clothes the patients wore had not been washed after their death. Without the knowledge that bacteria cause disease, people were not as attentive to washing as today.

- Those who washed more frequently, such as workers who handled mud and clay and other materials that they would want to get off their hands, did not get cholera.

- Nurses and doctors, who washed frequently, did not get cholera even though they worked with cholera patients.

This evidence suggested to Snow that the disease was spread not by the air but by liquid excretions from the body. Since these excretions normally went into the sewers, Snow then turned his attention to the distribution of disease and the distribution of water in the cities. The disease tended to be clustered, with some exceptions that caught Snow's attention:

- In the city of Manchester, those getting water from a well near a leaky sewage pipe got cholera.

- In Essex, there was an outbreak in one district served by a single well. A washerwoman living in that district was the only one who did not get cholera, but she used water from another well.

- In Locksbrook, a landlord who lived elsewhere was accused, during an outbreak, of providing poor water to his tenants. To prove that it was safe, he drank water being delivered to those buildings. He subsequently died of cholera.

With this information in hand he looked at a new epidemic in London. Ultimately 300 people died during the outbreak. Snow plotted, on a map, the residences of all the victims, and saw that they all clustered around one source of water, known as the Broad Street pump. A brewery nearby was not involved in the epidemic, but the brewers had their own source of water for the beer and did not use the pump. Six cases were in a different neighborhood but, when Snow got a map of the water pipes, he realized that the people in that neighborhood also got their water from Broad Street.

From this evidence Snow argued that the source of the epidemic was the pump. Furthermore, he argued, it was not a chemical contamination, since a chemical would be expected to dilute out with time and thus cause less disease; but the severity of the epidemic was continuing, suggesting that the cause could reproduce. Therefore the cause was likely to be biological, in other words, a germ. With that information, he finally did his experiment. He removed the handle from the Broad Street pump, rendering it inoperable. People in the neighborhood had to go to the pumps in the surrounding neighborhoods to get their water. Within a few days the epidemic was over.

Note what he proved and what he did not prove. He demonstrated that it was likely that the contamination came from one pump, that it was carried by the water, and that it was biological in origin. He did not identify the organism. In fact, the germ that causes cholera is extremely difficult to grow in the laboratory and, though widespread, does not often cause cholera. But its existence is one of the reasons we chlorinate water. Snow did not prove anything, in the sense that he had no true experiment to falsify his hypothesis. He could not intentionally force people to drink what he surmised to be contaminated water. He falsified competing hypotheses, leaving his hypothesis standing. It was enough to signal him to intervene, and the success of his intervention convinced every one of at least the pragmatic value of sterilizing water. The location of the Broad Street Pump is now commemorated in London[12].

Goldberger

Another famous medical experiment demonstrated that a disease was not caused by an infectious agent but rather by a quality of food. In 1912, there was considerable concern about an ugly, deforming, and debilitating disease called pellagra.[13] The name derives from the destructive skin lesions that are the first symptoms of the disease, but advanced pellagra can produce dementia and even kill. It was relatively common among the most disadvantaged citizens, including orphans and prisoners, and it was feared by all as potentially contagious or a genetic defect. Both theories were based on the most exciting scientific discoveries of the times. Darwin's theory of natural selection made people worry about the inheritance of diseases or the existence of "inferior," less resistant genetic background and the possibility of either propagating through society (based on some misunderstanding of the way in which genes or mutations spread in a population—see BTW1[7]). The second source of worry was infection. In 1859, the same year that Origin of the Species was published, Pasteur had shown that bacteria do not generate spontaneously, and he had subsequently shown that many diseases were attributable to infectious agents.

People naturally feared infection. Thus victims of pellagra were shunned and feared. So the Surgeon General's office sent a young, highly observant, physician named Joseph Goldberger into the U.S. South to see if he could find the source of the disease. Strictly speaking, this was not a situation in which initial experiments were possible, but to Goldberger where and how frequently pellagra appeared did not strike him as the pattern of a communicable or genetic disease. (See the discussion about Sir John Snow above about what pattern of a communicable disease looks like.) So he first toured the South, taking extensive notes and looking for correlations that might suggest an origin for the disease. The first common factor that he identified among the sufferers of pellagra was that they uniformly had very poor diets. It was common in institutions such as prisons and orphanages to fill the inhabitants' stomachs with monotonous but very cheap meals such as cornbread, molasses, and pork fat (only). He also noted that, in prisons, prisoners got pellagra but the guards did not.

To a physician, the fact that the guards did not means that the disease is not contagious or spread by an organism through the air or water. Furthermore, the guards went home at night and had at least two meals per day of a more varied, higher quality selection. While today a problem with diet would be obvious to us, at the beginning of the 20th C the idea that food varied in quality was pretty much scoffed at, as was the idea that there could be in food trace amounts of essential elements or compounds such as vitamins. By 1915 Goldberger was convinced that food was the issue, and he began his experiments:

- Hypothesis genetic flaw: Pellagra derives from a genetic weakness, such that people in the lower (genetically weaker?) population develop pellagra (unlikely since it does not seem to be inherited).
- Hypothesis infectious disease: Pellagra is caused by an infectious agent that is passed from a sufferer to an otherwise healthy individual (unlikely since unaffected individuals can move among affected individuals without acquiring the disease).
- Hypothesis inferior food: Pellagra is caused by a toxin in some foods or the lack of something in some foods.
- Experiment to test third hypothesis: Offer prisoners a varied diet or the monotonous diet. Compare results. Have the same prisoners switch diets. Compare results.

Goldberger found a prison that had its own farm and fed its prisoners from the farm. These prisoners did not have pellagra. He arranged that they would be pardoned if they tried the monotonous diet. They lived among the other prisoners and were not segregated in any way. After a few months the prisoners on the monotonous diet began to show signs of pellagra, while others did not. When they reverted to the varied diet, the pellagra disappeared.

Conclusions:

- The same individuals could either develop pellagra or not, depending on whether they ate the monotonous diet or not. IF they ate the monotonous diet, they developed pellagra. If they ate the varied diet, they did not. ONLY IF they ate the monotonous diet did they develop pellagra.
- Since the same individuals could either fall victim to pellagra or be cured of it, the hypothesis of a GENETIC FLAW IS FALSIFIED.
- Since the individuals participating in the experiment lived among the others, the hypothesis of CONTAGION IS FALSIFIED.

Goldberger expanded these conclusions by further experiments, which could be considered remarkably courageous but today would be considered unethical and foolhardy, not to mention disgusting. He conducted "filth parties" in which he and seven of his colleagues tried to infect themselves with pellagra. As Walter Gratzer writes, they "injected themselves with blood from severely affected victims … rubbed secretions from their mucous sores into their nose and mouth, and after three days swallowed pellets consisting of the urine, feces and skin scabs from several diseased subjects." The importance of the experiments is that they did not contract pellagra.

- The hypothesis of CONTAGION IS DRAMATICALLY FALSIFIED in that the most extreme efforts to contaminate eight people failed to produce disease.
- Prediction: Since pellagra is caused by a poor diet, improving the diet will prevent or cure pellagra.
- He convinced prison wardens and heads of orphanages to include fruits and vegetables into the diets of their wards, and pellagra disappeared from these institutions.
- The hypothesis of INFERIOR FOOD IS SUPPORTED BUT NOT PROVED by demonstrating that the prediction is validated.

This remains not absolute proof, since something else such as a small amount of toxin in, say, pork fat, that accumulates only when massive amounts are eaten, might still be possible. In any case, the prediction does not differ in kind from what was already observed, and thus is not a very daring one. When the prediction is something wildly original and unexpected, such as the bending of light by the gravity of the sun (BTW1[7]), it is much more convincing. Goldberger's argument was also limited a bit by ethical considerations. Since he could not morally conduct the obvious more convincing converse experiment, to take, for instance, well- to-do, "genetically high ranking," children and feed them only fatback (dried and salted fat from the back of a hog) and cornbread, this arm of the experiment remained incomplete, and he had to argue from his cure of children in two orphanages, who were cured within weeks by the change of diet.

One would think that, with such a dramatic result, Goldberger would be heralded as a hero and that there would be immediate reform of institutional procedure. However, that is not the way that the social implications of science work. The problem was not that the experiment was incomplete. In hindsight, it was convincing enough. The difference was the vitamin niacin. Vitamins are typically small molecules that are used in the body in various ways. The water-soluble vitamins, including niacin, are typically catalysts that speed up important reactions in the body without being used up. They are therefore recycled and need to be replaced only when they are excreted or otherwise lost—in other words, they are needed in only very tiny amounts. (The daily requirement for niacin is about 15 mg/day, about 5/10,000 of one ounce.) This was an issue: no-one had any idea that there were such micronutrients in food, and chemists were not even capable of measuring them even if they had been known. Basically, Goldberger had no mechanism by which to generate his hypothesis. Remember that mechanism is an important part of any hypothesis. Eugenicists (who argued that what we would today call the human genome had to be protected from being degraded by allowing "inferior" races and individuals to mix with the "higher" types) clung to their position that, while food might be a triggering factor, the basic problem was a weaker genetic background in the lower classes. Since knowledge of infection was relatively new and, as the "new kid on the block," infection was by many thought to be the source of all disease, proponents of the hypothesis of an infectious agent persisted in their belief. This is why Goldberger tried such drastic, if colorful, means of proving his point.

Most societies and individuals fight to preserve old ideas in the face of new, disturbing, propositions. If the world fails to end on the day that a cult predicts its doom[14], the response of the cult members is more likely to be "God heard our prayers" than "Oh, how embarrassing. We were wrong".[15] Other social issues conspired against the Goldberger theory. As I noted, science was infatuated with the idea of infectious disease. When an idea is new, it explains everything. Only as one gets to understand the idea more do the nuances begin to creep in. Furthermore, a growing sense of the meaning of evolutionary theory was leading to the conclusion of the struggle for existence and the eventual conquest by the "stronger". The assumption that poverty and disease were the result of an unlucky draw in the genetic lottery was a much more comfortable (and cheaper and less guilt-inducing) assumption than the alternative conclusion, that poverty and disease were the result of poor social management and should therefore be addressed by government and taxes. It took 20 years for Goldberger's argument to be accepted. Beyond the issues of the discomfort of accepting an idea that implied moral or social failure or potentially higher taxes, it really was just too novel an idea that there was something present in apparently miniscule amounts in food that could cause disease. Besides, he was working in the US South, which was still licking its wounds from the Civil War and smarting under the abuses of the Carpetbaggers. The arguments of an immigrant Jewish Yankee from New York, that would force them to deal with the possibility that poverty itself could cause serious disease, was a bit much to swallow.

The logic and evidence of science does not always win, at least in the short run.

Truly novel ideas are quite routinely rejected even by the top scientists in the field, indicating that even the best and most thoughtful minds are not always open to alternative analyses. Claude Bernard, considered to be the father of modern physiology, warned his students in 1859, "When the fact that one encounters opposes the reigning theory, one must accept the fact and abandon the theory, even though the latter, supported by impressive names, is generally adopted." Many successful scientists have been known to begin a lecture by citing the rejection or disbelief their ideas originally met. A colorful and recent example is the Nobel Laureate Rosalind Yalow. She, along with co-worker Solomon Berson, developed the radioimmunoassay, a remarkably sensitive means of measuring proteins, including some hormones, in the blood. They used the technique to measure the amount of circulating insulin. The assay was at least 1000x more sensitive than previous assays. They now could measure insulin in a drop of blood, and take repeated samples from the same animal, rather than, for instance, collecting all the blood from a dog or the pooled blood from several rats for one measurement. They found that, contrary to everyone's belief, insulin levels fluctuated rapidly, increasing perhaps 20- to 50-fold in a few minutes following a sugar meal and decaying to the original levels in about 30 minutes. This was truly startling and basically unbelievable. As the editor noted, "The experts in this field have been particularly emphatic in rejecting your positive statement..." because it contradicted then current theory. To be totally fair, however, the rejection letter received by Yalow did emphasize the conviction of the reviewers that the conclusions were not sufficiently justified— meaning that the reviewers wanted to see more definitive and unequivocal experiments. The experiments had been well done, but the results were so different from what everyone else had seen that, to the reviewers, they were unbelievable.

Perhaps the worst rejection letter of all time was that received by Antonie van Leeuwenhoek, the builder of the first microscope and discoverer of the world of microbes. He was totally amazed to realize that there were organisms smaller than one could see, and his amazement and wonder shines through his writings. For instance, when he scraped some plaque off his teeth and looked at it through his microscope, he described the experience like this:

> "I then most always saw, with great wonder, that in the said matter there were many very little living animalcules, very prettily a-moving. The biggest sort had a very strong and swift motion, and shot through the water like a pike does through the water; mostly these were of small numbers."

Or, in pond water,

> "In structure these little animals were fashioned like a bell, and at the round opening they made such a stir, that the particles in the water thereabout were set in motion thereby...And though I must have seen quite 20 of these little animals on their long tails alongside one another very gently moving, with outstretched bodies and straightened-out tails; yet in an instant, as it were, they pulled their bodies and their tails together, and no sooner had they contracted their bodies and tails, than they began to stick their tails out again very leisurely, and stayed thus some time continuing their gentle motion: which sight I found mightily diverting."

This was 1675. No one dreamed that there were creatures too small to see and that could cause disease. Malaria ("bad air") was recognized to be common in swampy areas, but was not associated with either the mosquitoes that carried the disease or the (still unseen) parasites that actually cause the disease. People consequently obviously took no precautions in the interest of hygiene. The bell-shaped creature with a tail would appear to be the protozoan Stentor[16], (also see a phase microscopic video[17] that illustrates very well the movements that van Leeuwenhoek observed) usually found in dirty pond water. Also, van Leeuwenhoek was a very skilled craftsman, producing far more perfect lenses than others who were experimenting along the same lines. The members of the Royal Academy of London, who reviewed the manuscript, could not believe that these creatures existed:

> "When I observed for the first time in the year 1675 very tiny and numerous little animals in the water, and I announced this in a letter to the Royal Society in London, nor in England nor in France one could accept my discovery, and so one still does in Germany, as I have been informed."

In a letter, Hendrik Oldenburg, the Secretary of the Royal Society, London, wrote the following to him:

Antoni Van Leeuwenhoek, Delft, Holland, 20th of October, 1676:

> "Dear Mr. thony van Leeuwenhoek, Your letter of October 10th has been received here with amusement. Your account of myriad 'little animals' seen swimming in rainwater, with the aid of your so-called 'microscope,' caused the members of the society considerable merriment when read at our most recent meeting. Your novel descriptions of the sundry anatomies and occupations of these invisible creatures led one member to imagine that your 'rainwater' might have contained an ample portion of distilled spirits—imbibed by the investigator. Another member raised a glass of clear water and exclaimed, 'Behold, the Africk of Leeuwenhoek.' For myself, I withhold judgment as to the sobriety of your observations and the veracity of your instrument. However, a vote having been taken

among the members–accompanied I regret to inform you, by
considerable giggling–it has been decided not to publish your
communication in the Proceedings of this esteemed society. However,
all here wish your 'little animals' health, prodigality and good husbandry
by their ingenious 'discoverer"

I, like presumably all scientists, have received many rejection letters in my time,
but in none of those was I accused of being drunk when I did the experiments. The
reviewers who rejected Yalow's paper at least argued that they wanted to see
more convincing evidence to overthrow a substantial body of work. The members
of the Royal Society did not consider it worth their while to attempt to falsify
Leeuwenhoek's hypothesis by asking, for instance, for evidence that distortions or
reflections through the glass could not produce such impressions. (Although the
properties of lenses were not well understood, this would not have been
unreasonable. Many a microscopist has seen a reflection of his or her eyelashes in
the middle of the field of view.) Actually, acceptance of the existence of
microscopic life and its effect on humankind has been a source of some
contention and can serve as a demonstration of the careful design of
experiments, since several times in the history of science there have been
vigorous disputes, including even prize contests, and the scientific world has been
convinced only following very careful demonstrations. From the 17th through the
early 20th Centuries learned folk have denied the existence of unseen and
dangerous organisms, sometimes for reasons other than logic. Ignaz Semmelweis
in the mid-19th C, before disease-causing germs were known, associated childbed
fever with a failure of obstetricians to wash their hands between leaving an
autopsy and attending a birth. He recommended a chlorine rinse, later modified
to carbolic acid (phenol) and, in doing so, cut childbed fever deaths from
approximately 15/month to approximately 2/month in his hospital in Vienna.
Such a recommendation however carried the implication that other obstetricians,
by being unclean, were in fact carrying the disease to the women giving birth. His
publication was met not with huzzahs but with antagonism, so much so that his
later life was a matter of constant embroilment in arguments. His theory was
much more readily accepted in the US, not so much because of greater insight in
the US, but because physicians on this side of the Atlantic felt less threatened by
the implication that European physicians were insufficiently cleanly.

THE CONTROL EXPERIMENT

A frustration among practicing scientists is the confusion, in the lay public,
between correlation and causality. For instance, one often hears a news item of
the order of, "I heard an airplane overhead, and a few minutes later my child on
his bicycle ran into a tree," implying that the airplane caused the accident. The
example is ridiculous, but it makes the point: just because two processes are
associated in time or space, it does not prove that one caused the other. A more

typical newscast would suggest that, because someone was exposed to a particular situation, he or she developed a disease. The recent and still extant flap about mercuric compounds in vaccines causing autism is an example, as is discussed earlier (p 12). Although signs of autism are often first observed when the baby's behavior develops enough for it to become responsive to adults, and this is typically near the age at which babies are first vaccinated, there is no evidence or mechanism to suggest that vaccination has any connection to autism. In fact, since encouraging parents to refuse vaccinations for their children creates a danger in itself, the physician who promulgated the argument was eventually censured for refusing to acknowledge that his argument and data were flawed and that the data of others were overwhelmingly against his claim.

How can we assert that the claim that vaccination (or, more specifically, that preservatives in the vaccines) cause autism was false? Scientists have two basic approaches to evaluate data: the control experiment, and statistics. Statistics is discussed above (p. 15). First, let's look at the control experiment.

The essence of any experiment is that there should be one and only one variable, which the scientist can control. When the scientist changes the amount of this variable, he or she should be able to detect a more-or-less proportional response. If I add more weight to a spring, the spring should stretch further. If I offer food to a baby animal, ranging from nothing to a reasonable amount, the baby should grow proportionately to the amount of food offered. The devil is in the details. For instance, late in the 19th C and early in the 20th C, much was learned about the function of our organs by surgically removing them from animals and observing what happened. However, surgery itself is a trauma that causes secretion of all stress hormones, which may also affect the response. The only way to deal with this is to cause similar trauma to another animal, without removing the organ, to confirm that it is the removal of the organ itself that causes the change. The scientific literature abounds with reports in which unrecognized other factors produced the results of the experiment. A night watchman's footsteps can, in the absence of other cues, allow animals to establish a daily rhythm; animals that have been laboratory-reared for hundreds of generations can show seasonal variation, because the quality of the grains used to make their food varies with season; some animals are more curious than others and come forward in the cage to see what is going on, and therefore create a biased selection; or one mouse can be frightened by hearing another mouse squeal. Sometimes it is extremely difficult to be certain that all conditions are exactly the same between the experimental situation and the control situation. It is sobering to realize that, even in mice so inbred that all the mice are more similar to each other than identical twins, they still manifest a range of lifespans. So, even if we conduct an experiment in which we consider all things to be equal other than the variable that we control, there may yet be something that we have not identified.

Often the issue is what is not yet known. I remember with some embarrassment a paper I wrote as an undergraduate. I was writing about the thymus gland, the function of which was unknown. I had come across a paper in which a French scientist, in the late 1940's or early 1950's had removed the thymus from baby guinea pigs. The guinea pigs appeared to do well at first, but later became cachectic (showed a wasting disease) and died. I wrote, with some pomposity, that the experiment had been done in postwar France, in which animal husbandry was probably not a high priority, and that if he had cared for the guinea pigs by the standards now (in the late 1950's) mandated by the U. S. government, he could have avoided the infections that his animals developed. I got an "A+" on the paper, from one of the period's leading immunologists. A few years later a series of discoveries from several laboratories demonstrated that the thymus gland was one of the centers of our immune system. The reason that the guinea pigs had died was because they had deficient immune systems. It is notable that even my professors missed the implication. As has been said, "There are no bad experiments, only bad interpretations."

We can also, without deliberately cheating, be influenced by our own bias. For many years, the structure of primate (ape) societies was described solely in terms of male dominance. There was an "alpha male", the one who is the natural leader and dominant member of the troop that led the tribe or pack, and all other members of the tribe followed him. There was plenty of evidence for the hypothesis, and little to contradict it. The fact that the hypothesis was developed by a German who had joined the Nazi party (though he later publicly regretted joining) and was used to justify some aspects of Nazi social theory left a bad taste in scientists' mouths, but it appeared to be valid. Early primatologists, who were all male, observed among apes the structure of male social hierarchy and interpreted the behavior of the females as totally dependent on the social ranking of their male partners. It was not until women primatologists, notably Diane Fossey and Jane Goodall, joined the research that the social structure of the females was noted and recognized to be important. Likewise, other primatologists began to notice that the "alpha male," was to some extent an artifact of zoos, where the animals do not have great variety in their lives. Field observations, combined with a more questioning or at least different political viewpoint, revealed that, while one male might be dominant (or, depending on your viewpoint, a bully) in a zoo setting, in the field one male might be the alpha defender against attack, another the alpha fruit-seeker, another the alpha chooser of the nesting site, etc.

Another, famous, presumptive case of bias was that of Mendel himself, the father of genetics. Many scientists have tried to replicate his experiments with peas. Though they get results similar to those he reported (for instance, getting three yellow peas for each green pea), their numbers are never as close to the ideal 3:1 ratio as Mendel's results. What happened? Well, peas can be yellow or green, but sometimes there is a pea in which the green color is not so intense, so that it is rather yellowish-green. One can speculate that, as Mendel began to recognize the patterns of inheritance, he began to look for them. Thus, when he came across yellowish-green peas, he subconsciously classified them as yellow or green according to how his count was developing; if he was getting fewer yellow peas than he expected, he might have decided that a yellowish-green pea was actually yellow.

Nobody accuses Mendel of cheating. It's just the way that humans are. For instance, if a doctor gives a patient a sugar pill without describing it, the patient is likely to report feeling better. This is called the placebo effect. If the doctor does not know that the pill is a sugar pill, the doctor as well might observe an improvement. This is why we often rely heavily on instruments to give us data, rather than relying on our judgment.

We also do blind experiments and even double-blind experiments. If I am trying to assess something that is not easily measured by instrumentation, for instance particular patterns seen under a microscope, it would be unwise of me to do the measurements entirely by myself. If some cells are larger and some are smaller and the experiment is supposed to change the ratios, I might unconsciously classify intermediate-sized cells according to my expectations. I would be much better advised to assign each sample a random number and then to ask a technician, graduate student, or someone else with enough expertise to make the judgment to count and classify each sample. Only if this person does not know which is which can I be certain of limiting bias. I should also set up a matrix or pattern such that the cells to be measured cannot be chosen by the observer, who might unconsciously choose the larger or smaller cells.

It is also essential that all the experiments be done at the same time. If the controls are done first, then the experiment, then all sorts of other factors might affect the experiment. Animals and some reactions can differ according to time of day; the water used to prepare the reagents might be different from one batch to another, or the reagents themselves could vary; the amount of light or temperature, which can vary, could be different between experiments; holding one set of samples while waiting for the second phase of the experiment to finish could affect the samples, which might deteriorate or otherwise change during the holding time; *etc*. There are an infinite number of ways in which an experiment can go wrong, and we have to do our best to account for as many as we can.

Reliance on machinery, however, has its problems as well. Our machinery is truly marvelous: In the 50 years since I was a graduate student, our sensitivity and ability to measure has increased at least one billion fold. We can now measure changes that occur within less than one minute in a single cell—something that was inconceivable when I was a student. In some circumstances, we can detect a single molecule!

However, with this sensitivity comes a greater responsibility, since we can now write papers and discourse at length on something that we never see. We rely on a readout from a machine to tell us that something has happened. While in most instances this works, a careful scientist runs many tests to assure that the machine under the conditions of an experiment is properly calibrated and does not produce false negative or false positive results. For instance, many of our experiments today rely on the very weak fluorescence emitted by signaling molecules. We need to demonstrate that our machines do not pick up stray light in the room or spontaneous fluorescence from other materials. We need also to verify that the fluorescence is not "quenched" or absorbed by something else (which can be turbidity, colored sample or solution, too alkaline or too acid conditions, or simply another molecule capturing the energy) before it reaches the sensors. It is a laborious task and one often subject to short-cuts by less demanding scientists. One of my preferred stunts in training graduate students is to create a deliberate distortion in the readout and let the student suffer (for a limited time) in trying to recognize and trace the source of error. If they found it, in congratulating them I would note, "Just because the machine cost $20,000 and gives a digital readout does not mean that it is right." As a member of editorial boards of journals, I have seen many manuscripts in which the numerical values presented for pictorial data presented in the paper could not have been correct, indicating that the machine was improperly calibrated or that the researcher set the parameters incorrectly, failed to account for limitations of the machine, or failed to address an artifact. In the curmudgeonly manner of older scientists, I note that in the days in which instruments were mechanical rather than electronic, calibrations were never stable and the machines had to be recalibrated every time that they were used.

We also must go to extraordinary lengths to avoid contamination. With the PCR (polymerase chain reaction, p. 113 ff) techniques so beloved by CSI-type programs, it is possible to amplify DNA 1,000 to 1,000,000,000-fold and identify, for instance, an individual from a single sample collected on a 10-year-old microscope slide, but it is also possible to amplify contaminant DNA in the assay. In early iterations of a technique now commonly used to separate and identify proteins (acrylamide gel electrophoresis, for the cognoscenti) researchers were puzzled by the frequent appearance of one protein that they had not yet identified. It turned out to be the common skin protein keratin, picked up from fingerprints containing dead skin cells when scientists handled with their bare hands the glass plates in which the experiments were run. We now require everyone handling the plates to wear gloves.

~~~~~

Chapter 3: Controls and verification of experiments

Adequate controls are the heart of science. Editors of journals ask their referees to look carefully at the quality of controls described in submitted papers. The controls are both "negative" and "positive". The negative control contains all ingredients except the one that is supposed to give a signal, such as a fluorescent compound. The purpose of the negative control is to assure that (in this instance) a truly dark situation gives no signal, that there is no light leaking into the apparatus, material that spontaneously fluoresces, or electrical or electronic quirk that causes the machine to give a false reading. The positive control contains a known amount of material such as a fluor, or a sample that is known to work in other situations (for instance, an enzyme known to be active in mice but being tested in this experiment in humans) to verify that the output of the machine accurately reflects the activity of the sample—here, that the light is not quenched or otherwise blocked, and that the sensors are working correctly.

One year, I taught a graduate course on experimental design and had the students read a series of papers of experiments that led to Nobel prizes. What we all (my students and I) learned from this exercise was that a number of prizes were awarded to scientists who were suspicious of the adequacy of their controls and investigated further. For instance, Joseph Goldstein and Michael Brown were trying to document, as had others, that cholesterol had no effect on cells in culture. However, unlike others, they were concerned that cholesterol is extremely insoluble in water, and decided to verify that it actually got into the cells that they were studying. Using a traceable radioactive form of cholesterol, they looked for it in the cells and could not find it there. They then assumed that it had bound to the walls of their petri dishes, as fats often do, but they did not find it there either. Finally, when they looked at the media in which they grew the cells, they found that the insoluble cholesterol was in fact dissolved in the media. Investigating further, they realized that it was bound to and carried by proteins: a lot of it loosely bound and easily pulled off of proteins called low density lipoproteins (LDL, the low density because all of the cholesterol—fat—that they were carrying) and some tightly bound to high density lipoproteins (HDL). LDL is "bad cholesterol" because the cholesterol can be easily pulled off and deposited elsewhere, more or less like water sopped up by a paper towel, while HDL is "good cholesterol," holding onto the cholesterol and preventing it from being deposited. Their discovery of the cholesterol-binding proteins led to their prize. The point here is that they did not conclude, "it has no effect," but they asked whether the experiment had actually been carried out as they thought.

The next year's prize was awarded to Stanley Cohen and Rita Levi-Montalcini. Dr. Levi-Montalcini had worked with Viktor Hamburger on a material that makes embryonic nerves grow. In embryos, peripheral tissues secrete this material, known as nerve growth factor, which attracts and stimulates nerves to grow out from the central nervous system and provide the innervation for them. However, since the experimenters could recognize the existence of the factor only by seeing the growth effect in embryonic chicks, they could not do much with it. They decided to try to determine its character by a process of elimination. If they could destroy the activity by digesting the extract with a protease (most enzymes, or biological catalysts, are named by describing the substrate they attack, with the suffix "-ase", hence "protease" = "enzyme that attacks proteins"), it would likely be a protein; if by digesting with a DNase, DNA; and if by digesting with an RNase, RNA. When they digested the extract with an RNase, they got good growth of neurons. So they concluded that the active factor was not RNA. However, as a good control, they also added to the culture of neurons the RNase itself, without the extract of growth factor. The RNase was a relatively impure preparation originally collected from snake venom. To their surprise, the RNase itself stimulated the neurons to grow! Enlisting the chemical skills of Cohen, they quickly determined that snake venom itself was a rich source of nerve growth factor. Working from this information, they were able to collect enough nerve growth factor to identify it—it was a protein—and purify it for experimental use. They thus identified the first of what is now known to be an extensive list of factors that stimulate the growth of specific cells, leading eventually to our ability to grow embryonic stem cells, potential directions to stimulate regeneration of lost or damaged organs, and many other important biomedical advances.

One of the most spectacular demonstrations of the importance of close attention to every detail of an experiment, and of the necessity to consider all alternative hypotheses, surfaced in 1973 and was described by Michael Gold in a book published in 1986[18]. In the midst of the War on Cancer, a major consideration and subject of interest was that normal cells typically did not grow well or indefinitely in culture, whereas cancer cells did. Furthermore, normal cells could sometimes be observed to undergo "crisis"; they would be gradually dwindling as the culture failed, when they would suddenly perk up and start growing very well. Furthermore, these newly-revived cells exhibited many of the characteristics of cancerous cells. There was considerable effort to determine what happened during crisis. Also, many researchers were convinced that cancers were caused by viruses (some are) and they searched for cancer-causing viruses. In a cooperative effort of the war on cancer, Russia sent some cells that apparently had a cancer-causing virus to U.S. government laboratories. As part of the analysis of these cells, one researcher, Walter Nelson-Rees[19], did some routine checks (that others were not doing). To his surprise, all the cultures that the Russians had sent carried a variant of an enzyme that was typically found only in people of African origin and, in the US, in only one out of three American Blacks. Also, ALL of the cells had markers that indicated that they came from women. Although the Russians had not given complete descriptions of the people from whom the cells had originally been collected, the population distribution of the Soviet Union made it extremely unlikely that all of the cultures that the Russians had produced would have come from women of African origin. Following a suggestion from Stanley Gartler's earlier work, Nelson-Rees began to piece together the possibility that the Russian cultures were all contaminated with HeLa cells, an extremely malignant cervical cancer that had been isolated and grown from an American Black woman named Henrietta Lacks and was the first human cell line that was considered "immortal".

This of course threatened to turn into a politically difficult situation, and Nelson-Rees decided to investigate further. In furtherance of the War on Cancer, many scientists had isolated many types of cells that had gone through crisis and were now being used as cell types representing many different types of cancers--of the mammary gland, prostate, thyroid, stomach, *etc.* He decided to check these. To his surprise and horror, many, including some of the most popular lines, proved to be female, and to carry the enzyme variant found in Blacks. They were all HeLa cells. Somehow, the cultures had become contaminated, and the very aggressively growing HeLa cells had overwhelmed whatever other cells were present and taken over the cultures.

We now know what happened. Disposable plastic pipettes and petri dishes were not common at the time, and we used glass pipettes, which had to be washed in devices called pipette washers. Circulation of water in these pipette washers was not always perfect, and in some laboratories the pipettes and dishes were not always carefully sterilized. If you overpack a pipette washer or an autoclave (to sterilize material) circulation will be poor and the material in the center, or a clogged pipette, will not be effectively washed or sterilized. Even in laboratories in which the pipettes were handled correctly, to get accurate measurements it was common (and appropriate) to "blow out" the last drop of liquid that remained in the pipette once it has drained. This was a trick that students learned in chemistry classes to assure accurate and repeatable volume of material delivered. Unfortunately, we now know, blowing out a pipette produces aerosols, microscopic droplets of liquid that float in the air. HeLa cells can survive in these aerosols and land in a different petri dish. (Often, while doing multiple transfers, scientists and technicians had many dishes open at the same time.) Thus, whether from poor washing, reuse of pipettes between one dish and another, or aerosols, HeLa cells had survived to contaminate many cultures and to lead the culture out of the presumed crisis. The problem was that most of the scientists, excited to have produced a new cell line with which to study cancer (and which might acquire the name of the discoverer or otherwise win him or her fame) did not look closely enough at their cells to note that they looked different or invest much effort in confirming that the cells were indeed variants of the original culture and not something else. As Nelson-Rees noted, perhaps one third of all cultures being studied by scientists were in reality HeLa cells. Moral: One has to be one's most harsh critic, to assume that even the most exciting finding is wrong, and to test even the most outlandish and ridiculous alternative interpretations.

As recounted by Gold, the story has a sad, if human and predicable, outcome. To learn that one's pet cells, for which one has invested considerable effort, and acquired some renown and substantial grant funds (past and future) to unlock their secrets, are contaminant HeLa cells, and therefore no renown and sharply diminished prospects of future grant funds, can ruin one's day. As in the case of Semmelweis (p. 35), Nelson-Rees' findings were received with a range of emotions that ranged from disbelief to skepticism to outrage. Though eventually the import of his findings was acknowledged, the bitterness of the controversy drove him from active research.

It still happens. Scientists generally freely share cell cultures carrying interesting genes or other characteristics. Infrequent, but not unheard of, is the situation in which the receiving laboratory attempts to verify the culture and finds something different from the claims of the shipping laboratory. Hopefully, today these differences are resolved amicably.

The tools of the trade: Control experiments

Many unforeseen things can affect an even very well planned experiment. One famous case involved an effort to select two strains of rats: a "smart" strain and a "stupid" strain. The way that the experiment was done was to continually select the smartest rats and separately the rats that performed most poorly on tests and breed each group separately. At the end of the experiment, it turned out that both strains had improved in their maze taking ability. What had happened was that, when the experimenters opened the cages, the most curious and most fearless rats came forward to see what was going on, while the more timid rats hovered at the back of the cage. The experimenters inadvertently always took the more curious rats whether they selected for high skills or low skills and so they finally selected for higher performance in both strains. In another situation, a group of experimenters wanted to find out what would happen to the diurnal rhythm (the rhythm that causes us to feel sleepy or awake at different times of day) if mice were kept in rooms with 24 hour light and no signals as to what time of day it was. Surprisingly, the mice kept perfect rhythms. It later turned out that, in the absence of any other cues, the mice synchronized their rhythm to the heavy footsteps of a night watchman who came through every night at a specific time. There are many such examples. If one injects a substance into a mouse, the act of injecting it is a serious stress to the mouse and may affect the results. The other mice in the cage may hear the screams and objections of the mouse being injected and may themselves become frightened and stressed. Some characteristics of a newborn mouse depend on whether, in the uterus, it resided next to brothers or sisters. More recently, some scientists have pointed out that cells cultured from nearly identical animals differ according to whether the donor was male or female.[20] There are so many potential variables in an experiment that it is truly a challenging task to identify and deal with them all.

Scientists always look at how the control experiments were done. The goal is that one has a set of preparations that are identical to the experimental preparations in every possible way except for what one has done in the experiment. If one injects something into an animal, one carries out an injection of a neutral substance in the control animal. If one adds a particular reagent to a solution, one adds an equivalent in amount of the solvent in which the reagent was carried.

The tools of the trade: Experimental Design

If a control is adequate then one is prepared to test the hypothesis. Remember that a hypothesis is a mechanism that is proposed to explain the relationship between something that is considered to be a cause and something that is considered to be the effect. Without a hypothesis, simply collecting data serves no purpose. All of the following comments have been made by working scientists.

- "Never do an experiment for which you have not planned a figure or table." (A figure [more specifically, a graph] or table shows the relationship between an experimental situation–the variable that one is manipulating–and a response that is purported to be caused by that variable, in other words, it demonstrates the validity of the hypothesis that is the basis of the experiment.)
- "Is this experiment hypothesis-driven?"
- "This is a shotgun experiment."
- "This is a fishing expedition."
- "Does this experiment have a sufficiently biochemical or mechanistic approach that it merits publication?"
- "This research is too descriptive."

In other words, scientists demand that research be directed toward testing hypotheses. Merely collecting data in the hope that something will fall out is disdained.

The essence of an experiment is that it tests a hypothesis—that is, a potential explanation of how one process called a cause elicits another process called a result. It is an explanation of how things work. The challenge is to ask a question in such a way that it produces an unambiguous result, such that the answer is either "yes" or "no" rather than "maybe". The experiment will never "prove" something; it can only eliminate plausible alternatives. Thus the hypothesis that warmth causes plants to grow is not well tested by observing plant growth in summer versus winter. So many factors differ between summer and winter— temperature, day length, intensity of light, amount of rainfall, the presence of insects—that we cannot unambiguously rule out these other factors and conclude that temperature is the only controlling factor. We must limit as many variables as possible and construct an experiment in which temperature is the only known variable. For instance, if we confine plants to opaque rooms in which the amount of light and day length are constant, and the amount of watering is similarly controlled, so that we vary only the temperature, we may observe a correlation between temperature and growth that we can interpret as causal. We will want to do the experiment in such a manner that two or more temperatures are run simultaneously, on the same batch of seeds, since if we were to run the experiments successively (first one temperature, then another) we would introduce many other possible variables. These might include a different propensity of plants to grow at different seasons, influence on growth by residues left in the soil or in the air from the first set of experiments, volatile factors from plants or other materials that vary with seasons, microscopic or microbial elements in the soil that vary with season or change during the first experiment so that they are different during the second run, etc. All of these are possible and in fact can occur. In essence, we need to work very hard to establish that the only known difference between the two experiments is temperature. Even so, there may be factors that relate to temperature that may be truly causal and mislead us into believing that temperature is the controlling factor. For instance, the evaporation of water differs according to temperature. If we choose an amount of water suitable for plants at the lower temperature, we might maintain the plants at the higher temperature under too-dry conditions, effectively dehydrating them and thus stunting their growth.

As I said, proper experimental design is not always simple or obvious. Perhaps the best way to illustrate this is to review the classical experiments described above that are remembered because of the clarity in which they made major arguments, along with more modern experiments that were distinguished by their questioning of the validity of the control experiment, as are described in the following chapters.

Testing a hypothesis

I can hypothesize anything that I wish. For instance, I can argue that the position of the stars will determine whether or not I will win the lottery, but to a scientist this hypothesis fails many tests. First, it does not propose a mechanism through which the stars can manipulate or constrain fate. One must include in the hypothesis a means by which some energy created by the position of the star can be transmitted both to my brain so that I choose specific numbers and to the balls being swirled around in the lottery drawing. For science works only with what can be detected and measured, or sometimes, as with the Higgs boson, something that is predicted to be detectable and measureable. Forces of energy or control that have never been detected or measured, like the influence of planetary configurations, are pure speculation, not within the realm of science. Cassius: "The fault, dear Brutus, is not in our stars, But in ourselves, that we are underlings." [21]

The second major failure of this speculation is that, by lacking a postulated mechanism, it does not contain within itself the suggestion of how to test it. A hypothesis has no value in science unless it can be tested, and the best hypotheses are those that, by being described, project specific tests that can be proposed. The tests may not be feasible at the moment. Sensitivity of measurement, for instance, may not be adequate, but the hypothesis suggests what needs to be developed and therefore spurs on science. Thus good hypotheses carry within themselves the possibility of their own defeat. One of the purest examples of such a situation was Einstein's hypothesis in 1916 that energy (here light) was a special form of matter, or that matter was a special form of energy, and that they were interconvertible. This shocking idea carried within it the seed of a potential test: If his hypothesis and calculations were correct, light like matter should be subject to gravitational pull. Flying objects are pulled to earth by gravity; and so should light. But how on earth (pun intended) would you test this? By Einstein's calculations, the effect would be infinitesimal. A bullet shot parallel to the earth will eventually fall to earth under the influence of gravity, but the effect of gravity on light would be MUCH less. A bullet may weigh a few grams and travel at ~1000 m/sec (~2,000 mph) but light travels at 300,000,000 m/sec and has unmeasurable mass. The answer was to measure the pull of gravity on light not on earth but outside of it. According to Einstein's calculations, the sun was sufficiently massive, and it would take light sufficiently long time to pass the sun, that one might be able to detect a light beam being pulled toward the sun, or "bent" by the sun. However, there is a problem with this experiment: since the sun is the brightest thing that we know (or knew at that time) how would the light beam be detected against the blinding light of the sun? Under one circumstance: during a full eclipse of the sun. This then was the proposed experiment: To watch a star passing behind the sun during a total eclipse. While the star was behind the sun but approaching coming out, if its light were bent, we would see the star appearing ahead of itself. Since our eyes and brain have no means of interpreting bent light, and would interpret the light as coming in a straight path, the star would appear to pop ahead of its expected position and then, as it finally cleared the sun, drift back into its appropriate place. On May 29, 1919 there was a total eclipse of the sun, an event eagerly awaited by the scientific community and much of the educated public. When the gravitational pull of the sun on light was confirmed, almost precisely according to Einstein's calculations, it made headlines throughout the world[22] (see animation[23]).

~~~~~

Chapter 4: Hypotheses: The Origin of AIDS

Polio and poplar trees

Hypotheses, like syllogisms, can be true or false. More problematic, though, is that, for a hypothesis, the conclusion may be true even though the hypothesis is false. The general structure of a hypothesis is, "If A causes B, then when I have A, I will get B." We could otherwise describe this as an "if and only if" situation: "If I have A, then I have B; if I DO NOT have A, then I do not have B." The problem arises when we recognize the correlation of A and B and presume a causal structure without testing it. For instance, during the polio epidemics of the 1940's and 1950's, someone noted a correlation between poplar trees in yards and incidence of polio. There could have been many possible valid links—for instance, if poplar trees attracted mosquitoes or other insects that carried polio—but it seemed unlikely that poplar trees caused polio, or that conversely poliovirus caused poplar trees to grow (an equally valid interpretation of the correlation). Subsequent study revealed the explanation that polio was a disease of the middle class, and poplar trees were found in middle class yards, as is illustrated by the table below:

Economic Class	Housing	Susceptibility to Polio
Lower	Small, crowded lots or multi-dwelling structures; no yards or trees	Children, living outside in the summer, are exposed to polio as infants, when the disease is rarely fatal but frequently quite mild and undiagnosed; they grow up immune to polio
Middle	Postwar housing on small lots planted with cheap, fast-growing trees like poplars	Children protected and clean during infancy, not exposed to polio; as children, they go to swimming pools, movies, or other crowded areas where contagion can spread; get polio as children
Upper	Large houses on spacious lots, trees never bulldozed or lots are planted with large, elegant, slow-growing trees	Children protected and clean during infancy, sent to private camps in isolated areas during summer outbreaks of polio; not exposed

Figure 4.1. How two unrelated variables, both of which correlate to a third variable, give the appearance of correlating. Here, middle class people were likely to have poplar trees in their yards and to live in a manner that exposed their children to polio. However, there was no causal connection between poplar trees and polio.

Therefore, a scientist must admit that a correlation suggests a cause-effect relationship that will form the basis of a hypothesis as to how it works. This will be a test of the "if and only if" relationship. However, it is always possible to get a true result even if the hypothesis is false. ("The warming of the earth causes the sun to rise higher in the sky and the day to get longer" is false but day length and overall warmth will correlate.) Therefore the goal of the scientist is to disprove or "falsify" the hypothesis. Here, for instance, one might attempt to find situations in which the earth got warmer and day length did not increase, or in which the earth got colder and day length did not decrease.

This is all very speculative and even silly. It helps to have a real example. One is to consider a quite plausible hypothesis that was put forward, that vaccination against polio caused the AIDS epidemic. This would obviously be a topic of extreme controversy and implication. We can therefore review how the scientific community examined the hypothesis, tested it, and ultimately falsified it. In doing so, we can examine how alternative hypotheses survived and therefore are still considered to be plausible. Finally, we can look at a recent denouement, in which an amplification of one hypothesis has introduced a new consideration and warning. This most recent step illustrates another characteristic vital to science: a hypothesis stands until disproven, meaning that there is no certainty. (As I tell students, to be a scientist one needs to simultaneously abhor, and have great tolerance for, ambiguity. The ambiguity that one must tolerate is present in every preliminary experiment, while the abhorrence provides the drive to try again.) If at any time a hypothesis can be convincingly disproven, then it is simply back to the drawing board. If, time and time again, using different techniques and methods (more about that later) we get only results consistent with the hypothesis, we rename the hypothesis a theory. It still might one day be disproven, but the likelihood is markedly less. Finally, if it is so unlikely to be disproven that no one would believe a contrary result (if a falling ball suddenly stopped in mid-air), it becomes a law.

To understand the story of Human Immunodeficiency Virus (HIV, the causative agent of AIDS) we need to know a bit about the virus and about the history of the time in which it appeared. HIV is a type of virus called a retrovirus. Most living organisms store their genes in DNA, and use the DNA to make a similar molecule called RNA, which then moves away from the DNA to participate in the construction of the proteins that make up most of our bodies. This process is described more fully in Chapter 7 (p. 83). Retroviruses exist containing only RNA as the genetic material. When they infect cells, the RNA is used to make a copy in DNA, which then remakes the RNA and eventually the protein of the virus. The "retro" in the name refers to the reverse step, RNA>DNA, rather than the conventional DNA>RNA. The problem is that the DNA>DNA replication, used when cells divide, is so basic a part of our biology that we have multiple means of ensuring the accuracy of the replication, destroying imperfect DNA, and repairing mistakes. The result is that there are very, very few errors. When RNA makes DNA, however, the machinery is nowhere nearly so precise, and there are relatively many more errors or, as we would call them in DNA, mutations. Retroviruses mutate rather rapidly, so that, as our immune systems acquire the capability of destroying one version of the virus, another one pops up, and, in AIDS, the virus keeps going until it destroys the immune system altogether. For flu, the high mutation rate explains why there are different types of flu every year.

We can also use these mutations to help us trace where and when HIV arose, and how it spread through the world. We can do this using a simple logical trick. Suppose, as seems likely, all humans originally had black hair and brown eyes. We also find a population that has blond hair and brown eyes. Among the blonds, we find some with blue eyes, but no blue-eyed, black-haired individuals. We might infer that the blond hair mutation appeared first, and that the blue-eye mutation arose within the blond population. If these populations are geographically restricted, the inference becomes even more likely.

Similarly, we can trace the ancestry of HIV through the accumulation of mutations and related techniques. We can even get a rough time scale. If, for instance, we know that, on average, mutations such as blond hair and blue eyes spread through populations only about once every 10,000 years, then we could estimate that the black-haired brown-eyed people have not shared a common ancestor with the blond-haired blue-eyed people for the last 20,000 years.

This now takes us to HIV. The most common form of HIV is rather similar to a virus that causes a mononucleosis-like infection of chimpanzees and is therefore called SIV (simian immunodeficiency virus) and to other viruses in various monkeys and apes. (Apes are the tailless animals most similar to us, such as gorillas, orangutans, chimpanzees, and gibbons, whereas monkeys usually have tails and are less likely to sit upright or stand.) So we can start with the possibility that HIV came from apes or monkeys.

AIDS (Acquired Immune Deficiency Syndrome) was first recognized among the homosexual population of the United States, from where it was quickly traced to identical, previously unnamed, disease in Haiti and in west-central Africa. This was the first basis of an intelligent but ultimately false hypothesis of its origin. The second component was a vaccine against poliomyelitis, a paralytic, highly contagious disease eradicated in the United States by 1979. Polio infected almost 60,000 people in the US in 1952. The vaccine was introduced in 1955. By 1956 the number of cases was down by half, and by 1963 it was virtually zero[24].

The relationship to AIDS begins with the development of the vaccine. Antibiotics work against bacteria but not against viruses. To combat viruses one needs vaccines. Vaccines were well known in the 1950's. Commonly, a live but heavily mutated and therefore not virulent, virus is injected, causing a very mild disease, and the infection stimulates the immune system to make antibodies that will defend against future infections of the unchanged incapacitating virus. This is the way that smallpox vaccination works (or worked; smallpox has been eradicated from the earth by vaccinating everybody). As William Jenner realized, farm workers who got cowpox (a very mild form of smallpox) did not get smallpox. He infected people with cowpox (vaccinia, from the Latin word for cow) and showed that they also did not get smallpox. Cowpox is the same virus, but very weakened. However, the infection stimulated immunity against both forms of the virus. The other way to create a vaccine is to kill the virus, but so gently that it is not too distorted to stimulate antibody production. This technique works if one does not have weakened or attenuated virus, though it is less effective than live-virus vaccine, since the injected virus does not reproduce and thus continue to stimulate the immune system.

Prior to the 1950's, there was no way to grow poliovirus in the lab other than to infect a monkey and hope to collect virus before the monkey died. Finally, John Enders and his colleagues at Harvard showed that poliovirus could grow in cells isolated from the kidney of a monkey, actually a species called a green monkey, a common monkey in West Africa. With this resource, many laboratories could grow large amounts of virus, and it became a matter of time until someone developed an attenuated virus or killed viruses in such a way that they would be suitable for use in vaccination. Finally, the US government put its money behind two leading laboratories, one to produce a killed virus (the Salk vaccine) and one to produce an attenuated virus (the Sabin vaccine). The Salk vaccine reached the market first, followed shortly by the Sabin vaccine, which now represents the primary form of the vaccine.

Hypotheses concerning the origin of AIDS: Part 1

There was a third contender for development of a vaccine, Hillary Koprowski. Koprowski was also developing an attenuated virus vaccine. He had influential contacts at the United Nations and persuaded this body to test his vaccine as well. It was tested in Belgium, Poland, and what was then known as the Belgian Congo. The Koprowski vaccine thus became an element of a hypothesis as to the origin of AIDS.

In the early 1990's writers in the US and Australia, realizing that most scientists considered that AIDS had apparently first appeared in or near the former Belgian Congo, pointed out that this was the region where Koprowski's vaccine had been used. These writers, Louis Pascal in Australia and Tom Curtis in the US, tried to publish their findings in scientific forums but their manuscripts were rejected or ignored. Finally, Curtis described his hypothesis in *Rolling Stone*, from which it was picked up by the *Amsterdam News* and other publications, gradually attracting increasing attention until Edward Hooper expanded it into a provocative book entitled *The River*. Briefly, the hypothesis was as follows:

- Koprowski collected green monkey kidney cells from a monkey that unfortunately was infected with SIV. (Such a situation was theoretically possible. In the 1950's, scientists did not know that viruses could be carried undetected in cultures of cells. We now know that viruses do not necessarily run rampant and kill every cell in sight; they can grow at the same rate that the cells do in a "latent" or "occult" fashion, expanding rapidly and killing cells only when the cells are stressed or in other special conditions.)

- When he infected the cells with poliovirus in order to mutate poliovirus and grow enough for use in vaccines, he inadvertently grew the SIV alongside the poliovirus.
- Thus, when children were vaccinated in central Africa in the late 1950's, they received both the attenuated poliovirus and a tag-along SIV.
- SIV, now in humans, became HIV. The children became sick but, because Congo erupted in violent civil war by the early 1960's, the illness of the children was overlooked or not documented.
- By the time that Congo became calmer, the disease was established and poised to spread throughout the world.

Obviously, such a hypothesis carries enormous political freight and is also a serious issue to consider for future attempts to immunize populations. There was at least some plausibility to the hypothesis, and it had to be examined. The scientific approach is to identify as many inferences as possible from the hypothesis—that is, to extrapolate every possibility suggested by the hypothesis—and test to see if the result is predicted. If the prediction proves accurate, the hypothesis is substantiated but, more reliably, if the prediction fails, the hypothesis is falsified and can be rejected. As we noted previously, one can get a true result with a false hypothesis, but the failure to meet a prediction dooms the hypothesis unless one can find an error or other factor that explains the failure.

The Pascal-Curtis-Hooper hypothesis carries several dubious elements, some of which are *de facto* failures of inferences:

- If Koprowski's green monkey was contaminated with SIV, one might expect Sabin also to have contaminated monkeys, but there was no evidence that AIDS followed vaccination with Sabin vaccine. The Sabin and Koprowski vaccines were given orally while the Salk vaccine was injected.
- If Koprowski's green monkey was contaminated with SIV, one might expect that his vaccine lots would have contaminated children in Poland and Belgium as well, but there was no evidence for an appearance of AIDS in Europe.
- The time scale, from vaccination in the late 1950's until the appearance of AIDS in the 1980's was a bit long, considering that the time from infection to serious disease was 5 to 7 years at the beginning of the epidemic.
- It is doubtful that, even in a country torn by civil war, one would not notice children dying of a severe wasting disease.
- The children who were vaccinated were young enough that few of them would have been involved in sexual activities before they became very sick.
- As we would later learn, if HIV came from green monkey SIV, then the genes of HIV should be very similar to those of SIV.

- The Koprowski vaccine was given orally, but, as we now know, HIV is very poorly transmitted through the digestive tract.

This is a very long string of "ifs," which can be considered to be probabilities, *i.e.*, the probability that Koprowski got a contaminated monkey but Sabin did not would be roughly inversely proportional to the frequency with which wild monkeys captured for laboratory tests were infected. The problem with probabilities is that they operate by multiplication: if I have a 50% chance (1 out of 2) chance of getting heads when I flip a dime and a 50% chance of getting heads when I flip a quarter, the chance of getting both heads is ½ x ½ or ¼, 25%. If a lottery consists of six numbers, each number being between 1 and 30, the chance of getting all six numbers correct would be 1/30 x 1/30 x 1/30 x 1/30 x 1/30 x 1/30 or 1/729,000,000 (0.00000000137 chance). So the more improbable or unlikely scenarios we collect, the less likely the hypothesis.

However, improbability is not refutation. Therefore, an intense search was done to examine further inferences.

- Inference: There would have been no AIDS before the introduction of polio vaccine. An intense search was undertaken of blood samples taken from people dying of mysterious diseases in the hope that someday there would be an explanation. Although there was a little uncertainty involved, a few samples appeared to be true HIV infections as early as the 1920's through the early 1950's.
- Inference: There ought to have been some, perhaps abortive, infection among recipients of Sabin vaccine and children vaccinated in Europe. None could be found.
- Inference: HIV should be very closely related to SIV from green monkeys, and much more closely related than to any other similar ape virus.

Although at first the technology was a bit crude, it became quickly apparent that HIV was most closely related to a chimpanzee virus, and that a second, less common form of HIV found on the west coast of Africa was most closely related to a virus from sooty mangabey monkeys. The green monkey virus was rather distantly related. Although there were also chimpanzees in Koprowski's laboratory, he did not use them to prepare vaccines. Furthermore, these chimpanzees had come from East Central Africa, whereas the chimpanzee SIV was found only in chimps from West Central Africa. Thus the hypothesis was in serious danger of being falsified or ruled out. The final nail in the coffin came when researchers located the remnants of the original Koprowski stocks. (Such is the sanctity of data and protocols that they are preserved until long after the scientists have retired.) These stocks were checked carefully.

- There was no sign of any type of SIV in the cultures.

- Finally, techniques had improved to such an extent that it became possible to trace not only the lineage of HIV but also the time when it most likely branched from its ancestral chimpanzee form.
- The differences between HIV and its closest relatives, chimpanzee and mangabey SIV, suggest that HIV became distinct from SIV as early as 1920 and certainly not later than the late 1940's or 1950.Thus the Pascal-Curtis-Hooper hypothesis bites the dust, as was argued in a public debate at the Royal Academy of London in 2000.

The death of one hypothesis leads to the erection or examination of alternative hypotheses, a process that will continue until no competing hypotheses remain standing, and even then the surviving hypothesis remains subject to challenge by new evidence. So we can now look at other explanations for the origin of HIV and AIDS.

Hypotheses concerning the origin of AIDS: Part 2

If we can now assume that the virus has appeared in the past, we can change the question to asking why AIDS did not appear earlier. One can look at the distribution of the viruses and their variants throughout the world, noting also when and where AIDS was first seen in various countries, to create a map of its spread. And one can look at social conditions and habits to see what they reveal.

It seemed clear that AIDS first appeared in Africa, and spread from there to Haiti, and from Haiti to the United States, first among the homosexual community. Thus a logical model of what happened was developed and considered probable until approximately 2005:

- Humans interact frequently with chimpanzees in Africa. Hunters hunt and butcher them, likely cutting or scratching themselves in the process. An open wound would provide a means of transmission of the virus. The hunters are men, but the women take the fresh meat and cook it, also a process that might result in an open wound.
- Humans probably therefore picked up HIV many times, with a low but not vanishingly low probability over time.
- However, during most of this history, people lived in villages and were not especially promiscuous. Thus, the person who became infected might infect his or her partner, but both would become sick in relatively short order, and would be isolated or shunned by the villagers and would die without infecting others.
- With the decolonization that occurred in the 1960's, the Belgian and French areas of the Congo improved roads and ports along the Congo River so that commerce could be maintained, especially since there were rich mineral resources inland that could be mined and sold.

- This improved communication system allowed the cities to grow, bringing workers into the colonial cities, which grew rapidly. Most of the migrants were men, creating very lopsided ratios of men to women. Thus prostitution spread rapidly.
- The prostitutes frequently played with wild or domesticated monkeys while waiting for clients, undoubtedly getting scratched or bitten on occasion. This would provide another means of transmission of the virus.
- The men would often stay in the cities for a week or a month or two, then return to their village to visit. If they had become infected, this travel would create a network whereby the virus could survive and spread.
- Gays often looked for vacation places in which, because of poverty or other social conditions, male prostitutes could be found. American gays vacationed on the west coast of Africa and in Haiti. They picked up the disease there and brought it to Haiti, from whence it spread to the US.

This story makes a lot of sense and is consistent with much that we know about history and social customs, but it is perhaps a bit simplistic and carries a certain overtone of blame, or prejudicial presumption about behaviors. Other regions of the world have become rapidly urbanized without concomitant outbreak of sexually transmissible diseases, and AIDS is overwhelmingly heterosexual in Africa and even Haiti, but, at least at first, most common among homosexuals in most of the New World and Europe. Finally, although HIV is contagious, it is not overwhelmingly so. The probability of transmission in a single sexual encounter is 1/10 or less. In fact, the variant HIV of West Africa, HIV-2, the one from sooty mangabeys, appears to be dying out. Fewer people become infected than die of the disease. Something is missing.

Hypotheses concerning the origin of AIDS: Part 3

In this context Jacques Pepin, a Canadian physician-researcher who participated in efforts to eradicate yaws (a syphilis-like disease) and malaria in central Africa, has added a new component and a different twist. He builds his argument from three observations:

- It is very difficult to get a disease started unless there is a high probability of transmission. If the probability of infecting a sexual partner is only once in ten contacts, then one infected man coming in from a village would have to have intercourse with ten prostitutes in order to, on average, infect one, and that prostitute (one among several hundred to a few thousand in a city) would have to have ten further contacts simply to maintain the existence of the infection. This might be possible in the cities, but it first presupposes a high rate of infection among the prostitutes, and there is no obvious mechanism to create this initial rate of infection.

- If, however, you have a very high rate of infection, then the probability of the infection becoming self-sustaining increases exponentially. In other words, if 1 of 100 people is infected, the probability of passing it on will be 1/100 x 1/100, or 1/10,000, but if 1 in 3 is infected, the probability becomes 1/3 x 1/3 or 1 in 9.
- In contrast to the low infectivity by sexual contact, transmission by injection is highly efficient. We know this well because hepatitis C is very easily transmitted by non-sterile needles or, before we could test for it, by infected blood, and several people were infected with HIV by being pricked with needles that had been used on patients with AIDS.
- In many parts of the world in the 1950's there were massive efforts to immunize people against various diseases. These efforts had value: they led to the complete eradication of smallpox and the drastic reduction of polio and many other diseases. However, in the 1950's there were no plastic disposable syringes. Syringes were glass and needles were washed and reused. In countries with elaborate medical resources, syringes and needles were carefully sterilized by autoclave (essentially a pressure cooker) between uses.

As Pepin notes, in Africa there were almost no autoclaves and needles were cleaned with bleach (not favored because it leaves toxic residues), alcohol, or formaldehyde, or sometimes even simply washed with clean water or saline solution. We know today that these procedures are inadequate to destroy viruses. In Haiti, Pepin notes, there was a company that collected blood for sale to the Red Cross and for clotting factors. This company used the same inadequate sterilization procedures, and many of the sellers were poor people who came repetitively to the collection center.

Thus Pepin adds a new layer to the hypothesis that HIV arose from a spontaneous transmission of SIV and jumped into humans many times, but that increasing urbanization became a conduit for its spread. He argues that something else is necessary, and that this something else is transmission by contaminated needles.

- Massive vaccination programs, coupled with insufficient sterilization of needles, created a reservoir of infection sufficiently large that it could become self-sustaining. (Rather than condemning these programs, he notes that they did eliminate or bring under control many hugely debilitating diseases.)
- The U.N. sent Haitian soldiers, who were Black and French-speaking, to the region of the Congo and Rwanda as peacekeepers in these war-torn countries. It was these mostly heterosexual soldiers who picked up HIV and brought it back to Haiti.

- The blood collection plant in Haiti became a means of expanding the reservoir of infection in Haiti. Among the infected were male prostitutes who passed the infection to US-based homosexuals.

Pepin's hypothesis fits much more of the now-available data, which traces the timing and routes of spread both by variations in the virus and by medical records, but it remains a hypothesis, that is, a proposed mechanism to explain the origin and spread of a frightening disease. It differs from the polio hypothesis in that, unlike vaccination against polio, these vaccinations were by needle, and the timing is more reasonable. The importance of the hypothesis is that it allows us to predict what might happen in a future effort to control disease, and to modify our procedures accordingly. If our hypothesis is true, we will avoid a future terrifying epidemic, primarily by using disposable syringes and needles.

This hypothesis, like its predecessors, is subject to continual extrapolation of inferences, testing, and potential eventual falsification. Our point is made: science is the continual floating of trial balloons, with our whole energy devoted to trying to shoot them down and float our own trial balloons. In this game, the biggest prizes go to two types: those who float balloons so clever, unforeseen, and logical than no one can shoot them down, and those who devise extremely clever means of shooting down false hypotheses. What is often very important is that the experimenter asks, "But is the control right? Can I trust that it is an adequate control?"

~~~~~

Chapter 5: Where ideas come from

Metaphors

Sometimes all it takes is a metaphor. As of this writing, the abstruse biomedical terms "programmed cell death" and "apoptosis" register over 230,000 different publications on the most thorough listing of biomedical publications of note, known in the field as PubMed, and almost 24,000 are being added every year (almost 3 every hour, 24 hours/day). Although the story is, as usual, much more complex and rich than even most scientists believe, in one telling it begins with a metaphor.

Let's take a minute to clarify what we are talking about. The idea behind programmed cell death is relatively simple and, in retrospect, obvious, though in one sense counter-intuitive: that, in our bodies cells don't just die, they most often commit suicide or are murdered. Apoptosis, a related idea, was based on the observation that most dying cells looked alike as they died and therefore the creators of the term posited a common mechanism of death. We'll explain these a bit more thoroughly in a historical context, but it helps to know why there are so many publications in the field.

The importance of cell death

Both concepts have been validated, and it turns out that proper regulation of how and when cells die is a fundamental part of our biology and that many diseases or abnormalities derive from poor control over cell death. In embryos, virtually every time that tissues separate to form organs, the separation is achieved by the death of cells in the middle: when muscles separate to form the musculature of the outer body wall and the musculature of the inner organs; when holes or lumens (literally: light) open in organs such as secretory glands; when the growing posterior digestive tract meets the growing anterior digestive tract and the two fuse to form a continuous structure. When the two sides of the face grow together, cell death along the midline makes a sticky surface to which they adhere, creating the complete face. If it does not work properly, we get cleft palates, harelips, or worse. The separation of the little palette that will form the hand or the foot into fingers involves a very precisely timed cell death. As the nervous system and brain differentiate, the most efficient way to get everything wired correctly is not to have each nerve (for instance, from the retina of the eye) carry specific instructions as to where it is to go and connect—this would take an encyclopedia of information-carrying genes far larger than we have available in our chromosomes—but to vastly overproduce nerve cells, allow them to grow in a general direction and make contacts, and then to keep alive only those that make the most useful contacts, allowing the others to die. Up to 2/3 of the nerve cells that are born die in the embryo or shortly after birth. (This, incidentally, may be the reason why activities that interest infants are so important: during the time that all this is happening, a more active brain keeps more neurons alive.) The reason that we can make antibodies against almost every bug that we encounter, yet usually do not make antibodies against our own proteins, is that our body has an ingenious way of taking the precursors to antibody-producing cells and breaking up, mixing, and recombining the genes that make the antibodies. The result is that, before we are born, we have an almost infinite variety of potential antibody-producing cells. Then, in a highly orchestrated process, the cells that can potentially make antibodies against our own proteins, and therefore are already active, and the cells that make ridiculous products, and therefore are too inactive, are killed off, leaving only the cells that can sit in reserve, ready to be activated if a particular bug shows up (Fig. 5.1)

Limbs, organs, and hollows are sculpted

The CNS is Wired

The immune system is selective, but not against self

Figure 5.1. Cell death is an important aspect of development. Left: Upper: The digits on the hands and feet are sculpted by death of cells between them (blue cells on left figure). Lower: The openings of many organs and ducts are created by the death of the inner cells (blue). Similarly, all vertebrates begin to develop both male and female organs but, for each individual, one set of organs ultimately dies and is resorbed. Center: Many neurons are formed in the central nervous system and grow competitively to reach target organs. Only those that establish successful connections survive. In this case nerves coming from the retina ultimately establish contact with the visual cortex in the back of the brain. The arrows indicate the distribution of an image before the eye and on the visual cortex. Right: Of the many immunologically competent cells that are created (upper left) those that identify self die (red cells, upper right) as do those that do not recognize any type of molecule to which to respond (green and yellow cells), leaving only those likely to recognize a foreign bacterium or virus.

The process continues throughout life. In most tissues, cells are replaced throughout life. For instance, in the liver one cell in one thousand is replaced every day. We used to think, "Cells die and then are replaced," meaning that for some profoundly uninteresting reason a cell would die and the active, controlled part would be the body's detecting that death and causing another cell to divide to replace it. What has changed is that we now consider that the death of the cell—how and why it dies—is also interesting or even more interesting. When the process fails, we have disease. To give two examples: (1) What kills in Acquired Immune Deficiency Syndrome (AIDS) is not the virus, Human Immunodeficiency Virus or HIV, but an overwhelming infection that results because there are not enough immune-system cells to fight it. The cells are gone because they have died. What is a bit crazy is that most of the cells that die are not heavily infected by virus. Rather, nearby cells that are infected send out a series of wrong signals, convincing these "bystander cells" to commit suicide rather than sit waiting for another infection or act against the virus itself. (2) We now know that cancer is often more a disease of failed cell death than of excess cell multiplication.

Three observations led to this new understanding of the dynamics of cancer. First was the realization that, although transplantation of organs such as kidneys undoubtedly saved the lives of patients, the transplant recipients (whose immune systems were suppressed to prevent rejection), and AIDS patients, developed more cancers than would normally be expected. Second, calculations based on good knowledge of the stability of DNA and the amount of radiation, from cosmic rays or naturally-occurring elements, led to the conclusion that humans ought to be forming two or three potential new cancers every day. Finally, scientists began to question why we had such a vigorous immune rejection of transplants (requiring suppression of the immune system to protect transplanted organs) when, never in our evolutionary history, was it necessary to reject tissue from another human being. These queries coalesced into a hypothesis of *immune surveillance*, which states that every human normally generates many potentially cancerous cells but that the immune system actively watches for them—they are marked by not having the right combinations of proteins on their outer surfaces, or, in other words, by being hopelessly unstylish in their clothing—and destroys them as they appear. In those unlucky souls who develop cancers the immune surveillance fails to identify or kill the offending cells. Suppressing the immune system, by virus or by chemotherapy, increases the probability of this failure.

Why does immune surveillance fail to kill potentially malignant cells?

The field takes off

An explanation came from two more-or-less contemporaneous observations, approximately in 1990, that arose from the new ability to identify and isolate specific genes, combined with a growing interest in cell death: two major genes associated with cancer had functions related to cell death. In the first case, a type of relatively slower-growing lymphoma named B-cell lymphoma proved to have an alteration in a gene that was normally inactive. In a case of a chromosome breaking and reattaching in the wrong location, the gene had been moved next to a promoter (p. 81) for a gene that was normally active, so that this gene, named Bcl-2, for B-cell lymphoma gene 2, was continuously active. (Promoters are sections of DNA generally next to structural genes [genes that ultimately produce proteins] that control the activity of the structural genes. See Chapter 6, p. 74.) The lymphoma cells did not proliferate especially more rapidly than normal, but the normal B cells typically lived only a few days. The mutated (lymphoma) cells persisted for much longer times, accumulating in huge numbers as their replacements kept piling in. In the second case, Bert Vogelstein and colleagues had identified a gene, named p53 (gene that produces a protein 53 kiloDaltons in size) that was mutated in most gastrointestinal cancers and more than half of all cancers.

Their first impression was that the role of p53 was to monitor the health of the chromosomes. If there was an abnormality on the chromosome, p53 would prevent the cell from dividing again until the abnormality was repaired. Thus a cell with mutated (damaged or absent) p53 would still be able to divide even if its chromosomes were abnormal, a frequent situation in cancer cells. However, Tyler Jacks and Scott Lowe realized that p53 did more than that. If the cell had a chromosomal abnormality but nevertheless had already started the preparations to divide (with the progress toward division quickly becoming irreversible) then the normal version of p53 would cause the cell to commit suicide before it completed the division. The mutated form of p53 could not provoke the suicide, and the cell might then progress to become a cancer.

Thus two major cancers were associated with abnormalities of cell death rather than abnormalities of proliferation. Cells are profoundly social, listening to signals from their neighbors and many other cells to tell them what to do. Today we know that the flaw in many cancers is their refusal to commit suicide when other cells are telling them to or their inability to commit suicide, and a massive research effort is directed toward influencing that decision. Even for those cancers that are not characterized by abnormalities in cell death, there is an effort to develop drugs that will select only the cancer cells ("target them") and trigger the cell death mechanism in them. Thus the study of cell death has become hugely important in cancer research.

Finally, in a third connection between cell death and cancer, nearly simultaneously Peter Krammer in Germany and Shigekazu Nagata in Japan identified a previously known gene, the protein product of which sat on the surface of the cell. When another protein interacted with this surface protein, the interaction killed the cell. It was originally identified as an anti-cancer mechanism— Krammer's results were spectacular in destroying a particular laboratory-designed tumor—and is now recognized as being both one of the ways that the body can destroy cancer cells and a major mechanism by which cells producing potentially anti-self antibodies are destroyed, as well as a mechanism by which excess antibody-producing cells are eliminated after an infection is resolved.

On the other side of the coin, many cells die when we don't want them to. Such cells include cells in the brain, which normally survive one's entire life from birth to death. Neurodegenerative diseases such as Alzheimer's Disease, Parkinson's Disease, Huntington's and Senile Chorea, and many others result from death of nerve cells. Although we are getting to know better what chemicals are toxic to these cells, it may not be possible to directly alter metabolism to prevent the accumulation of these chemicals. An alternative and complementary approach is to recognize that the cells that eventually die are agonizing for many years, as these chemicals gradually accumulate, and that their death is a suicide, a decision made after many years of suffering. In addition to alleviating the suffering of these cells, which we may or may not be able to do, we may be able to offer greater support so that we can delay or prevent the commitment to death. Neurodegenerative diseases are also a focus of studies of cell death.

Finally, even medical problems that seemed to be unrelated are now being investigated. For instance, a classic uninteresting form of cell death was infarct (a blockage of blood flow that produces a heart attack or stroke). Here the mechanism was obvious. The muscle or brain cells distal to (on the far side of) the blockage were starved of oxygen and nutrients and therefore died. However, again with knowledge of cell suicide, animal studies and clinical observations led to a different conclusion: following an infarct, heart muscle cells survive quite a long time, hours in fact. Even if blood flow can be re-established, many of the cells, grievously injured but not dead, commit suicide. In the laboratory, if this late death can be prevented, the size of the infarct or area of dead tissue can be cut to 1/3 or ½ of its potential size, with consequently much greater potential for the heart to recover and continue functioning well.

These laboratory findings are beginning to propagate into clinical approaches to treat patients. These are the practical reasons why the study of cell death has become such an active one in today's biomedical world. Now it is time to return to look at why the field lay fallow for many years before suddenly blossoming in the 1990's.

Origins of the field

In the 19th century, German dye chemists were exploring biological and mineral means of coloring garments. The results can be seen by a visit to most art museums. Women's dresses now appear in fuchsia, lavender, and chartreuse – colors that had not been seen before. The earlier microscopists, who were stymied by the transparency of cells and tissues, particularly when the tissues were cut into thin sections, appreciated the possibility. If a dye could stain wool, which was a protein, then it could stain proteins within the cells. If it could stain flax, linen, or cotton, then it could stain carbohydrates within cells. Thus histology, the science of studying cells tissues under the microscope, was born. Fairly quickly thereafter, these 19th-century scientists realized that cells divided to expand numbers and that occasionally they died. Given the *fin de siècle* attitude of the late 19th century, these deaths attracted some attention, particularly in the obvious instances of metamorphosing animals. However, though the existence of cell death as a biological phenomenon was noted, nothing could be done other than to observe it. By the mid-20th C cell death was an incidental aspect of development.

During and immediately after the Second World War, a more mechanistic approach was born. Dame Honor Fell in England, studying the differentiation of cartilage in culture, noted that the cells that formed cartilage died in the process of doing so. Rita Levi-Montalcini[25], working in the laboratory of Viktor Hamburger in St. Louis, tried to determine why there were more nerve cells in and along the spine where there were limbs than in regions, like the lower back, where there were not. By moving limbs into different locations in chick embryos, they discovered that limbs gave off a chemical, now termed Nerve Growth Factor, that stimulated the nerves to grow and, more importantly, not to die. They realized that far more nerves were born than survived, and that many more survived if nerve growth factor was present. Eventually Levi-Montalcini and Stanley Cohen won a Nobel Prize for isolating and identifying nerve growth factor.[26] Also, John Saunders at the University of Albany described small regions that predictably died in growing chick embryos. These regions of dying cells separated the limbs from the rest of the body wall and separated the toes of the chick feet. In birds that had webbing between their toes, such as ducks, there was far less cell death.

By the early 1950's Alfred Glücksmann, at Cambridge University, was able to compile a long list of instances in which cell death was a normal part of life. He considered that most deaths had to do with developmental or evolutionary reasons: getting rid of vestigial organs, sculpting limbs or organs, etc.

The experimental phase began a few years later. In one set of experiments, Saunders showed that if he dissected free the section of the chick limb that would be expected to die, and put it in a petri dish with all appropriate goodies and nutrients, it would survive for a while but die on schedule. One could conclude that the tissue was already moribund but that he had not recognized it, as for instance, an animal might seem to be OK while its kidneys are failing or a massive infection is building. However, he ruled out that possibility by putting the excised tissue not into a petri dish but by transplanting it into the back of another chick embryo. In this case the tissue healed in, survived, and became part of the back of the host chick. Saunders was able to conclude that the tissue had not committed to death at the time he excised it or, as he wrote in 1966, "the death clock was ticking". See animation of the experiment[27].

Our own experiments followed the same logic. I had joined the laboratory of Carroll M. Williams as a new doctoral student and he, in a manner common to training for doctoral research, suggested to me several problems that might interest me. This is a "not-all-eggs-in-one-basket" approach. If the first idea does not pan out, the second one might. In any case, one of the problems was the disappearance of muscles during the metamorphosis of insects. The larva of a butterfly or moth (caterpillar) is very different from the adult, and has many specialized structures that are not found in the adult. In this instance, all the many muscles that allow a caterpillar to hump along as it does are gone in the adult, which has its own set of muscles with which to fly, walk, or manipulate its abdomen. A few of the larval muscles persist in the pupa. If you have ever seen a pupa of a moth, it can twirl its abdomen when it is perturbed. When the moth emerges from its cocoon, these muscles serve to inflate the crumpled-up wings. By squeezing its abdomen vigorously, it pumps fluid into the wings, which unfurl like a Chinese snake. The wings then harden, and the moth flies away. Beyond this point, moths will never again have a protein-containing meal, and these remnant caterpillar muscles represent a last good piece of meat. The muscles are rapidly broken down and their resources reused. Of the many caterpillar tissues that are destroyed during metamorphosis, we chose these latter for a very practical reason: we could store the cocoons in a refrigerator all year long and use them for experiments when we wanted, rather than being forced to work only when caterpillars were spinning cocoons at the end of the summer. The question was, how are these tissues destroyed, and what initiates this destruction?

Figure 5.2. The original experimental evidence for programmed cell death. Left: A dissection of a caterpillar showing the first of three layers of muscles that fill each segment. All but those in the abdomen disappear when the caterpillar pupates. Right: the abdomen of a moth is opened along the dorsal (back) surface and the viscera are removed. The abdomen is then pinned, revealing the body wall with the mid-ventral line (belly) in the middle and the head toward the top. The abdomen on the left, from a moth just after it has emerged, is filled with the powerful muscles (the white fibers running from head to tail) that have persisted since the animal was a caterpillar. Two days later (right) almost all of the muscles have disappeared.

A couple of years earlier, a Belgian group headed by Christian de Duve had discovered intracellular particles that they had named lysosomes, assuming that the particles were responsible for killing cells. Lysosomes contained enzymes capable of digesting cells and, in carbon tetrachloride poisoning, the enzymes were released into the cells, damaging and destroying the cells.[28]

This release turned out to be an artifact of the manner in which carbon tetrachloride damages the cells, by dissolving its membranes, and we now believe that lysosomes are the equivalent of a digestive tract for the cells. Nevertheless, we didn't know that, and it seemed like a good starting point—did lysosomes kill the cells?

My investigation led to a much more complex understanding of what was happening. We identified hormonal and neural steps that had to occur before the muscles became primed to die, and the priming included the preliminary synthesis of lysosomes and, finally, the conversion of the lysosomes into a form in which they "ate" individual parts of the muscle.

Programming (moths)

- Hormones start development, including capacity of muscles to die
- Late in development, proteolytic enzymes increase
- Neuro/neuroendocrine activity at emergence of moth activates potential to die
- New enzymes are synthesized
- Muscles are non-functional by 12 h, digested by 24 h

Figure 5.3. The original evidence for programmed cell death.

Now we needed a title by which to describe it all, one that would sum up the major point of what was to be my doctoral thesis. Here's where the metaphor comes in. Williams was well known for his colorful and memorable way of describing things ("A moth is a flying machine devoted to sex!") and as students we tried to emulate him. The computer business was nascent at the time, and it was always a plus to be in the avant-garde. Thus, in the afternoon teas in which we discussed our results, we appropriated a computer term and began describing my results as programmed cell death. I am credited with coining it. Certainly it was the title of my thesis, but it arose during the teas and someone else may have first used it. We can't tell now. In any case it was a metaphor. Archibald McLeish, in a course that I had taken (as an undergraduate I was very involved with poetry) defined a metaphor as a means of making someone see something differently and went on to explain, more or less (my memory is a bit hazy) that poets saw something that everyone could see, but made them see it in a new light, whereas scientists saw something that no one had seen, and made them be able to see it. In that sense, "programmed cell death" was more like the poet's version. It seemed relatively obvious to me. If a characteristic appears at a particular stage of development on an animal and in a particular location, then—perhaps very indirectly—it is part of the developmental program encoded in the DNA. In the same sense that you can say that there are genes that determine the formation of limbs or the shape and color of eyes, you can say that the predictable death of muscles in a caterpillar is programmed into the genes. But the time was ripe, and the metaphor "programmed cell death" caught on. It helped other scientists to see cell death in a new light, as a fundamental biological event, interesting and worthy of study in its own right.[29]

There was now impetus to study the mechanisms of programmed cell death, but it was still a relatively quiet field in developmental biology. One other major event had to occur before, in the 1990's, interest skyrocketed. The other major event was to generalize the topic to "real" (mammalian, medically relevant) biology. In the 1960's John Kerr, a quiet, careful, and thoughtful pathologist in Australia, had been asking how cells died. The usual explanation was that cells expend most of their energy in getting rid of water. To be technical, they pump out ions and the water follows. Once they lost their sources of energy, for instance if their access to oxygen was cut off or no nutrients were available, water would still come in by osmosis, the cells would swell, and they would eventually burst. This happens all the time if we take an organ from the body, where it no longer has circulation, or even if we cut thin slices so that oxygen has a chance to diffuse in. However, many of the cells that Kerr recognized as dying did not look swollen. If anything, they were shrunken: condensed, with little blebs (bubbles) coming off of them, and with the nuclei shrunken and rounded. The DNA in the nucleus looked very peculiar, flattened against the nuclear membrane rather than diffusely scattered throughout the cell. He saw this type of cell in many situations. He described it as "shrinkage necrosis". Then, in 1966, he went on sabbatical leave to Scotland, visiting the laboratory of Alastair Currie and his graduate student Andrew Wyllie. Looking at the material from all three researchers, they concluded that "shrinkage necrosis" was a quite common form of death, seen in many types of cells and circumstances, and that it might even be by far the most common form of death. It obviously also could not be explained as an osmotic rupture. It had recently become routine to identify the rate of cell division in tissues and organs by introducing a label that would become incorporated into newly-synthesized DNA and—since that strand of DNA will persist for the lifetime of the cell—looking for that label, even years later. The three researchers decided that what they were looking at had to be a controlled process, the obverse of cell division (mitosis) and they searched for a term that would emphasize that death was a biological part of the life cycle of a cell. Consulting a scholar of Greek, they chose a term that described the falling of leaves from a tree or falling hair as one goes bald, and furthermore was similar enough in sound and construction to mitosis that the parallel would not be missed: apoptosis. As with "programmed cell death," "apoptosis" was a metaphor, though this time from the science side (seeing what has never been seen before), and it worked. More and more scientists reported seeing apoptotic cells.

An Apoptotic Cell

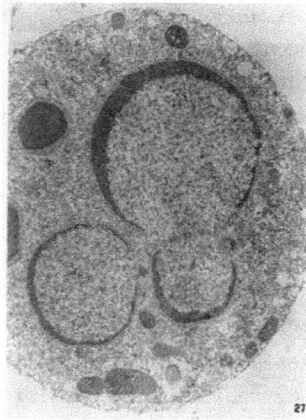

Figure 5.4. An apoptotic cell. The nucleus is breaking into fragments and the DNA (very dark material) has collected along the nuclear membrane. The cell has rounded and shrunk, while the other organelles in the cell seem to be undamaged. (They have changed, but the changes are not visible in this type of microscopy.) Wyllie, Kerr, and Currie correctly deduced that these changes could not be the result of loss of control of osmosis. They also emphasized that this form of death was seen in many different situations, and therefore could be defined as a biological phenomenon in its own right.

However, the mechanism of recognizing apoptosis was confined primarily to electron microscopy, a technique that involves very laborious techniques to examine, finally, a piece of tissue perhaps 0.002 x 0.002 x 0.00001 mm in size. We now know that the entire process of apoptosis can be completed in less than one hour and, unlike a cell that has undergone mitosis, an apoptotic cell cannot be labeled so that one could detect its previous existence after the fact. One researcher calculated that the entire liver could disappear in one month without a researcher seeing more than two or three dying cells in a standard microscope section (10 x 10 x 0.006 mm). The only thing that could be done was to observe its occurrence; it could not be readily studied. Something more was needed.

That something more was accomplished in four discoveries of the late 1980's and early 1990's. We have already encountered the first two. The realization that cell death was important in cancer, and that at least two major cancer genes were perhaps first and foremost cell death genes, stimulated interest to find out more about these genes and how they worked. Third, Wyllie's group determined that the strange appearance of DNA in these cells resulted from the DNA's being digested or broken down in a peculiar way. They could create lots of apoptotic cells by exposing cells from the thymus glands to cortisol, which kills them (the reason that cortisol suppresses inflammation, since these cells cause it). Pieces of DNA can be separated by running them through a gel. A gel, a jelly-like substance but stiffer, consists of a tangle of fibers of specific molecules, gelatin in the case of jelly, agar (a carbohydrate) in the agar that is used for petri dishes, and more complex molecules in the case of plastics. By making the gels in specific dilutions, one can get tangles of fibers such that the pores are just large enough to let molecules like DNA through. DNA has a negative charge and will move toward the positive pole of an electric field, the anode, and the smaller pieces will wiggle through faster than the larger pieces, separating the pieces of DNA by size[30]. Wyllie's group saw that, in apoptotic cells, the DNA was not broken down randomly, in which case the pieces of all possible sizes would produce a smear on the gel. Rather, it was broken down into specific pieces. DNA in its native form is wrapped around balls of protein, such that wrapping one ball takes a string about 180 base pairs long. What they saw was that the DNA was cut into pieces 180, 360, 540, 720.. base pairs long, as if it had been cut between the balls only (Fig. 5.5). One bacterium was known to be able to do this, but there were no bacteria present. The mechanism is interesting, but here the point is that separating DNA pieces like this (electrophoresis) is cheap and easy. It does not require very expensive equipment, and it can be done by almost any laboratory. What was important was that it provided a means by which any laboratory could look for apoptosis. Many laboratories did, and many reported that they, too, had detected apoptosis in all sorts of developmental, diseased, and normal states of tissues and organs. It was time to concede that apoptosis was a widespread and important biological process.

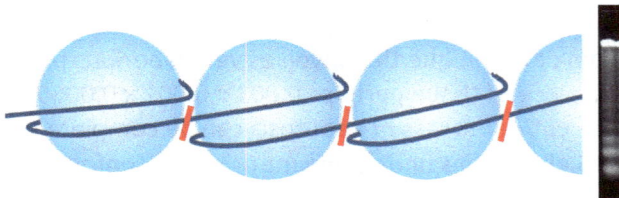

Figure 5.5. The structure of a chromosome. The "rope" represents the DNA double helix, which is wrapped twice around a complex of proteins called histones, here represented by balls. The enzyme cuts the DNA only where it is not in close contact with the histones (the red lines between the balls). Since not every location is cut, this leaves pieces of DNA one ball, two balls, three balls... in size. When the protein is digested away and the pieces of DNA are separated by size, the DNA separates into a discontinuous "ladder," illustrated by the image at the right. (The bright white band at the top is intact DNA.)

The final really important discovery was that of the genetics of cell death and, even more important, the implications of that genetics. It was the icing on the cake. In the 1970's Sydney Brenner, one of the gurus of the developing molecular biology, and a scientist whose way with words led to many aphorisms ("The question is whether development is American or British in style" [meaning, whether a cell differentiated according to what its heritage was (British) or according to its environment, or what its neighbors thought (American)]; "Development produces 'a sort of a worm'" [meaning: at the finest level, development is not totally specified but adaptable]) that helped define the attitudes of modern molecular and developmental biology, suggested that the only way to truly understand development was to know everything that one could know about a single organism. He chose a tiny, transparent, rapidly breeding roundworm that contained, as an adult, 1039 cells not counting eggs or sperm. Each worm had exactly the same number of cells and developed in exactly the same way. With his postdoctoral students John Sulston and H. Robert Horvitz, he set out to identify the lineage of every cell in the organism. As he told the Medical Research Council of England, he might not find anything for a while, but when he did, the story would be important. The U.S. National Institutes of Health would never have supported him on such a premise, but the MRC did, based on his reputation. (For the record, it is dubious whether the MRC would today.) The three set to their task and in 1975 published the complete family history of every cell in the organism. The implications of their observations led to several changes in the way that we look at how organisms develop. More germane to our story, the story of the lineage included 131 cells that were born only to die before dividing again or going on to do anything else.

C. elegans

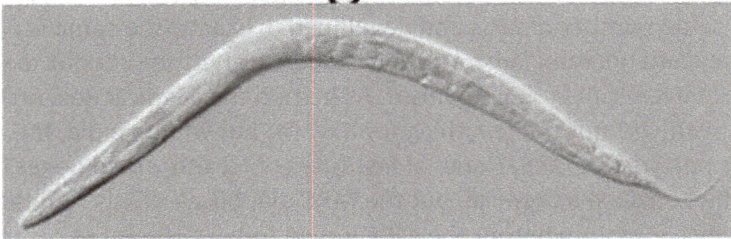

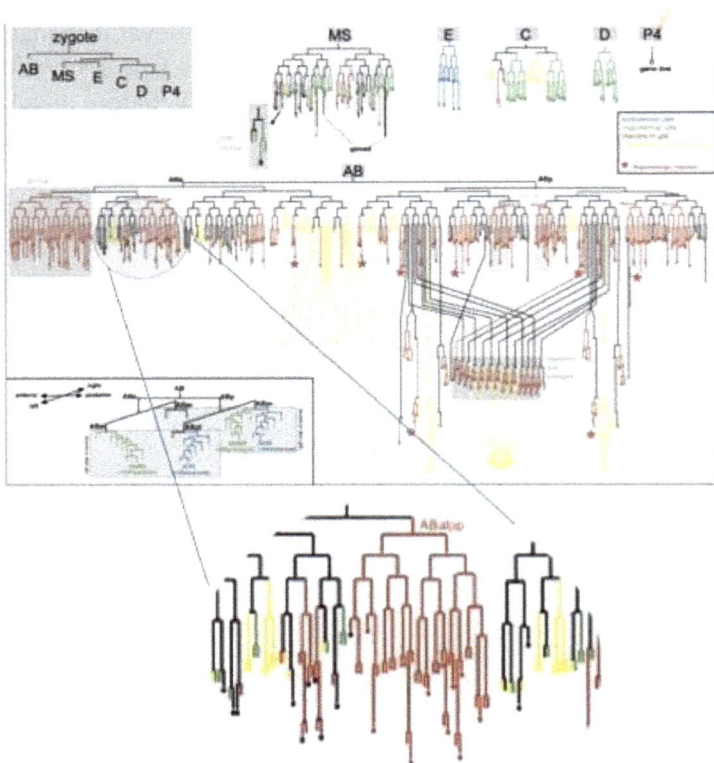

Figure 5.5. The history of the nematode worm *Caenorhabditis elegans*. Upper: a photograph of the worm, the actual size of which is approximately 1 mm. Lower: Genealogy of the cells of the worm, as described by Horvitz, Sulston, and Brenner. Each division of each cell is followed. Many of the cells divide only to have at least one of the daughter cells die immediately thereafter (black dots terminating lineages, easily seen in enlargement at bottom).

They also searched for genes that affected the development. Among the genes that they identified were a few (originally three, now known to be several) that affected *all* the deaths. Mutations in either of two of them would prevent all deaths, with the extra cells being incorporated, more or less harmlessly, into surrounding tissues. Mutations in a third would increase the number of deaths or kill the embryo. What was important here was that there were genes for cell death. The original presumption was confirmed. The situation is more complex in other animals, notably insects, and it would have been far more difficult to identify them in other animals, but then, that is the purpose of selecting simple animals as Brenner had proposed. The idea that cell death genes were real, rather than theoretical, was exciting on its own and a profound stimulus to explore further, but the story was to get even more exciting. Horvitz pursued the idea in his own laboratory at MIT. A graduate student, Junying Yuan, joined his laboratory and, now that the techniques had come far enough, she undertook to identify the cell death genes by sequencing them. The first one that she managed to sequence was breathtaking. It was the gene for a protease, an enzyme that digests proteins; this could explain how it killed the cell. More important, it closely resembled a protease known for humans.

When one finds genes that are very similar in extremely different organisms, it usually means that these genes are "conserved"—they are so important that anything other than very minor changes will kill the animal, and therefore that the gene evolves very slowly or not at all, because most mutations in this gene produce animals that do not survive. To find that a cell death gene in a roundworm was conserved in humans meant that it had to be at the center of something very important in humans. Subsequent studies have confirmed that the entire cell death mechanism is conserved from roundworms to humans, and we now know a great deal about it.[31] For instance, we know that in most instances cells are pre-programmed to die and that they normally are actively prevented from committing suicide. This is a fail-safe way of getting rid of cells that are in trouble by virtue of infection or other causes are potentially dangerous. It is rather like the behind-the-lines spy whose instructions are, that if he is ever not certain that the signals are completely correct, he is to swallow the poison pill. For AIDS and potentially cancerous cells, as well as for the gradual turnover of cells in most parts of the body, the correctness or incorrectness of the signals determines whether or not a cell will commit suicide. All cells are in some manner or another highly socialized, depending for their life on their neighbors' judgment.

Programming: C. elegans

- All cells contain the capacity to die: to be destroyed by the protease CED-3
- However, CED-3 must be activated by binding to CED-4, and CED-4 is held by CED-9. CED-9 is therefore a suppressor of cell death.
- At appropriate developmental times and in the appropriate cells, specification genes such as EGL-1 produce proteins that will bind to CED-9, releasing CED-4 to activate CED-3.
- Other genes are responsible for engulfment of the dying cell and destruction of its DNA

Figure 5.6. The structure of programmed cell death in the nematode, as worked out primarily by the group of H. Robert Horvitz

Programming: C. elegans

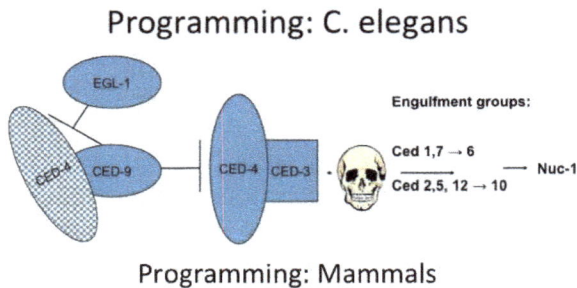

Programming: Mammals

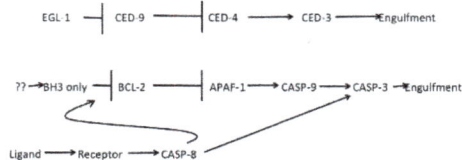

Figure 5.7. The structure of programmed cell death in worms (upper) and the similarity of the program between worms and mammals (lower; the upper is the nematode and the lower is for humans). In scientific convention, an arrow means that the action is driven in the direction of the arrow, while a line ending in a vertical bar indicates that the molecule to the left inhibits the activation of the molecule indicated to its right. Not only are the pathways extremely similar in overall structure, the molecules themselves are so similar that in many instances they may be substituted for each other.

With this armamentarium available—a "quick and dirty" means of identifying apoptosis, knowledge that lack of control of apoptosis is central to many disease states, and recognition of genes that activate, effect, or prevent cell death, research into cell death took off and is now the huge enterprise we see today. Nevertheless, although we are coming closer, there is as yet no wonder drug based on controlling cell death.

Why is that? Well, science is pretty much an onion. We peel off one layer only to find another layer, another question, underneath. There was huge excitement in working out the mechanisms of cell death. However, a complicating factor is that, once we got past the excitement, which included the certainty that apoptosis could account for all deaths,[32] it became apparent that some cells died not by apoptosis but by other mechanisms. More importantly, now that we have great understanding of the process and mechanisms of apoptosis, our ignorance becomes more visible. In many instances of disease the machinery of apoptosis is present and functional but either is activated when it does not need to be or fails to be activated when it should. The next layer of the onion is how apoptosis is turned on and off, and what signals regulate that decision. This is a considerably more complex question, most likely involving the current metabolism of a cell, its past history, its communications via local diffusion with its neighbors, its long-distance chemical communications with other parts of the body, and perhaps many other elements such as previous dietary decisions. Nevertheless, without this information, we are in the position of knowing all the players on the team and what their roles (offense, defense, particular activity for that position) are, but without knowing the rules of the game. Finally, we now know that, should apoptosis be blocked, there are alternative means by which cells can kill themselves. We will get there, it will come, but the politicians have it wrong. You cannot will an immediate cure to a disease. It takes a lot of minds, thinking in different directions, and sometimes specific techniques, questions, or intellectual ferment to break through. [Back to end of Chapter 9, p. 126]

Science is international

Since then there have been many brilliant advances in the field, mostly relating to how the proteolytic enzymes are activated by signals from other types of cells or by changes in how energy is accessed and used. In addition to the United States, contributing researchers have come from (in alphabetical order) Australia, Brazil, Chile, Canada, England, France, Germany, Hungary, Israel, Italy, Japan, the Netherlands, Singapore, Turkey, and several other countries. Many other countries have been the source of important information, if not directly then via expatriates working in other lands, originating from China, Iran, Jordan, Lebanon, Russia, and other countries. If one can describe one type of success in life as being able to feel comfortable in any society and with any people, then scientists progress through life with a huge advantage. When we meet and interact with scientists from other countries, we start with the bond of seeking answers to the same questions and the desire of seeking answers together. For instance, a recent paper, by a Japanese group publishing in a US journal, used reagents and programs purchased from suppliers in several states of the US, Russia, Germany, Ukraine, Canada, and Japan. These commonalities transcend our national differences, and it is not unusual to see close relationships develop between scientists whose governments are strongly opposed. If you ask scientists to name their closest friends, it is highly likely that they will name someone who lives in a different country. It is an unusual, if not unique, situation; most scientists consider the ramifications to be a source of pleasurable thought.

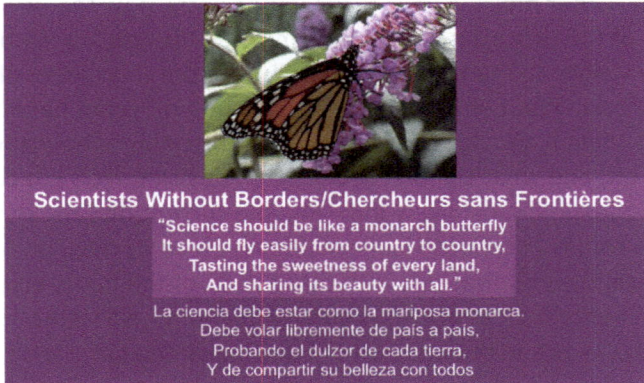

Figure 5.8. The motto of a group of scientists who travel to third-world countries to aid scientists and students in those countries

~~~~~

Chapter 6: Tools of the trade: Inventing technology

Scale

A particular characteristic of science is that questions hugely predate techniques. Some of the most sensational discoveries reflect answers to questions that have long been in the wind. On occasion, the answer comes from a uniquely brilliant insight, but often, especially in the biological sciences, it comes from the advent of a new technique. In the approximately 50 years since I received my Ph.D., our ability to measure things has increased over one billion fold. When I was a student, I was considered pretty good if I could measure properties in approximately one milligram of tissue. One milligram (10^{-3} g) is equivalent to a few crystals of sugar, 0.000035 oz. While this was good, we could not hope to understand how a few cells in an embryo, in the space of minutes or an hour or two, could commit themselves to become a brain, or a limb, or a liver. Today we can measure activities in materials measured in femtograms, or 10^{-15} g. What this means is that if we were to dissolve a sugar cube into a lake of water approximately 10 miles x 10 miles x 100 ft deep in size, we could still detect the sugar in a sample of water equal to the size of the original cube. We can watch how a single cell in as little as a few minutes can commit itself to becoming a particular type of cell, abandoning all other possibilities.

Warburg

But let's take a quick step back and look at how such sensitivity was achieved, looking at an ingenious mechanical device to measure and evaluate how oxygen was used and how foodstuffs were digested in animals. Some of the best work in science is done by the people who devise the machines and the techniques to resolve these old questions. One example is particularly illuminating, because it is a mechanical device, the function of which is relatively easy to understand. Otto Warburg, a member of a German banking family who became one of the fathers of biochemistry—he had such prominence that the Nazis could not touch him and instead declared him to be "an honorary Aryan"--built it so that he could answer the question of what tissues did with oxygen and how they turned the burning of oxygen into energy. His solution was an ingenious technique to measure chemical reactions when the chemistry itself was extremely limiting. He was trying to determine the differences between fermentation (conversion of sugar into lactic acid or, in yeast, alcohol without using oxygen) and respiration (conversion of sugar into carbon dioxide and water, using oxygen), and under what conditions each occurred. The basic reaction is as follows: in its simplest form, $C_6H_{12}O_6$ ➔ 2 $C_3H_6O_3$ (sugar in its simplest form, grape sugar or glucose—sucrose, the common table sugar, has 12 carbons) is converted to two molecules of lactic acid, yielding a little energy in fermentation. $C_6H_{12}O_6 + 6 O_2 \rightarrow 6 CO_2 + 6 H_2O$ (in the presence of oxygen, six oxygen molecules are used and six carbon dioxide molecules are produced for each molecule of sugar consumed, yielding approximately fifteen times more energy in the process called oxidative respiration or oxidative phosphorylation).Warburg was working out these now well-known reactions and was also trying to understand the conditions under which each occurred. He was also interested to know the differences between using carbohydrates like sugar and fats or proteins, both of which took more oxygen than sugar to produce the same amount of carbon dioxide. Finally, in his studies he realized that cancer cells more typically used fermentation rather than respiration, and he considered that this difference might be diagnostic and might provide an opportunity for treatment. (The distinction is still valid, owing primarily to the poor vascularization of tumors, and is considered in designing therapies.) His solution was ingenious. The chemistry was still primitive, but both oxygen and carbon dioxide are gases, and the volume of a gas can be measured. If you put a long straw in the opening of a balloon, and seal the balloon around it, then place a droplet of water in the straw, a relatively quite small change of volume of the balloon[33] will be reflected in a large excursion in the droplet. You can thus get a quite sensitive measurement of a small percentage change in volume. Furthermore, carbon dioxide will react quite readily with alkali, say sodium hydroxide or lye, forming sodium carbonate, which is a salt, not a gas. Therefore, if oxygen is used up, the droplet will retreat toward the balloon (reaction flask); if carbon dioxide is generated, the droplet will move outward. If oxygen is used and carbon dioxide is produced, the movement of the droplet will be more-or-less cancelled. However, if you can set up two identical preparations, one of which contains alkali to trap carbon dioxide and the other of which does not, in the former you will measure only the consumption of oxygen, whereas in the latter you will measure the net result of oxygen consumption and carbon dioxide generation. A simple subtraction will give you the amount of carbon dioxide produced. This was the secret of the Warburg flask. Using banks of these devices, Warburg was able, in a time when microchemistry did not really exist, to evaluate how respiration and fermentation worked and under what circumstances.

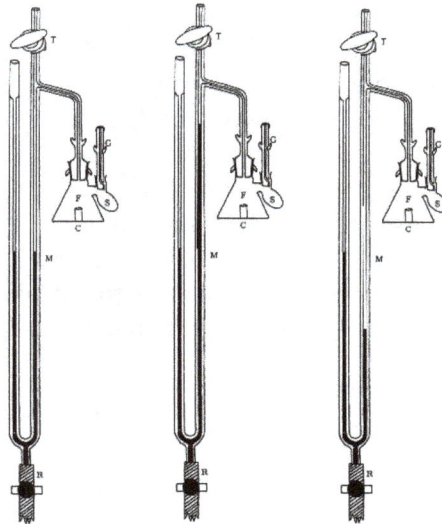

Figure 6.1. A Warburg manometer. Left: F: The reaction flask. S: a sidearm that contains substrate, which can be emptied into the reaction flask by tilting when the flask has adjusted to the temperature of the bath in which it sits. C: the container that can hold alkali to absorb carbon dioxide. M: The liquid in the manometer itself. If gas is consumed in the flask, the liquid will rise in the right arm; if gas is generated, it will fall. Middle: When the reaction consumes oxygen (CO_2 trapped by alkali), gas volume decreases and the fluid rises in the right tube. Right: When the reaction produces gas (fermentation, CO_2 generated without consumption of oxygen, no alkali present), the liquid falls.

Isotopes

One of the turning points was the ability to measure radioactivity. Prior to the time when we could, the means of measuring what was going on in the body consisted of attaching a colored molecule to another one and trying to trace what happened to the color as it passed through the body or through a tissue— relatively crude, and subject to debate as to whether the attachment changed how the molecule was handled, the extent to which the colored molecule itself was affected, etc.

Radioisotopic forms of molecules have almost no impact on biochemical reactions and they may be measured with extreme accuracy. Furthermore, the radioactivity of an individual atom is not altered by chemical changes that the molecule undergoes. It thus becomes possible to incorporate a radioactive atom (coming, since the Second World War, as waste material or by-products of nuclear reactors) into a molecule. This molecule can be delivered to the cell or animal and will be processed essentially normally, whether it is consumed or is incorporated into the structures of tissues, as it is processed in the body, and identified at intermediate or final stages. The atomic decay of a radioactive atom is measurable as radioactivity. Without too much difficulty, we can detect of the order of 30 decays/minute above background. Because of the huge numbers of individual atoms in a small amount of substance, this means that, depending on the specific isotope, in a single milliliter or gram of substance we can detect between 0.01 micrograms and 0.1 nanograms of isotope. (0.00000001 and 0.0000000001 grams; for comparison, a cube of sugar weighs about three grams). This was a huge advance in our capabilities, both because of the enormously increased sensitivity and because most of the isotopes did not immediately interfere with the biological processes. Prior procedures of analysis had consisted of feeding massively excessive amounts of substrates to be consumed or analogs of substrates containing colored markers that could be traced. Both of these techniques had some advantages but were likely to (and did) seriously alter the metabolism from normal, and thus produce many results that were clearly artifacts.

For technical reasons, however, it was often not possible to measure the radioactivity in a living cell or organism, so that one could assemble sequences only by using different organisms at different times. For instance, to trace the fate of a vitamin or a nutrient, one would have to introduce the radioactive material into several animals or groups of cells, and then at different times look for it in different animals or batches of cells. It was an enormous advance, but left a lot of questions still open. Besides, radioisotopes were expensive, they required expensive machines to detect them and special precautions to handle them, and it became extremely expensive to dispose of them. (In the early 1990's, when I was in charge of such an operation, it cost us $350 to dispose of a single oil drum containing all the loose trash that had come in contact with any radioisotope. Furthermore, we had to reserve a locked room to store it before it was picked up; we were overall handling less total radioactivity than was likely lost by a hospital that used radioisotopes for diagnostic purposes or therapy; and the price was increasing very rapidly.) Therefore there was an effort to find alternative procedures, procedures that would lower the cost, limit risk, and allow continuous monitoring of an experiment, preferably in single animals or cells.

These advances included two other primary directions. One was the ability to identify specific protein molecules in living or dead tissue, by using highly specific antibodies.

Antibodies

During the 1970's observations by several laboratories led to the conclusion that each antibody we made was made by a different type of cell. This does not mean an antibody against, for instance, measles. There are hundreds of them. An antibody recognizes, and binds to, a very small particular kink, fold, or other identifiable part of a protein, perhaps a unit involving four to six amino acids of the two to three hundred amino acids that make up a protein (range: 70 amino acids to a few tens of thousands). The measles virus itself contains several different proteins. A person immunized against measles will have hundreds of antibodies to each of these proteins, and each antibody will be produced by a single type of cell. Where these cells come from is a fascinating story in itself, but not relevant here. The point is that the specificity of these antibodies is such that some of them will recognize a feature that exists on only one protein of all the possible proteins, and thus this antibody will provide a remarkably specific and accurate means of identifying that protein. We use tests based on this principle to identify pregnancy, specific types of blood, and numerous disease states. However, there are three complications to prevent us from using the antibodies: First, to make a test specific, you have to isolate the specific antibody or the cell that makes it. Second, unfortunately the specific antibody-producing cell does not grow or multiply. It simply makes antibodies. Third, you have to identify the interaction between the antibody and its target, the antigen. The major issue was the second complication. This was resolved by an ingenious technique that involved the observations of several laboratories, but most notably that of César Milstein and Georges Köhler, who fused the antibody-producing cells to cells of the same type that had turned cancerous. They had realized that, in the disease multiple myeloma, the multiplying malignant cells produce antibodies. By fusing the antibody-producing cell in which they were interested to a cell of the myeloma that did not produce antibodies but could multiply, they created antibody-producing cells that could divide and grow, or hybridoma cells.

However, the process of producing hybridomas was random. You could isolate an individual fused cell and allow it and its descendants to grow, and thus produce a clone of antibody-producing cells, but now you had to know which antibody this clone was producing and whether the antibody was one that would be useful or not. To do this involved another trick. The interaction between antibody and antigen can in specific circumstances trigger a whole sequence of reactions that, in the body, serves to alert other cells to the presence of foreign material. The reactions will work in culture plates, and even the minutest reaction can be amplified by Enzyme Linked Immunosorbent Assay (ELISA), in which the initial reaction activates an enzyme that itself can chew through thousands of color-producing molecules, thus producing a color that can easily be detected. With this color reaction, one can now undertake a very but not impossibly laborious procedure. By growing up to a few thousand individual clones—one can get little dishes called multiwell plates that each have 96 separate wells in them, thus ten plates = 960 wells—and scanning for those very few that produce color, it is possible to find a clone producing a very specific antibody, or a monoclonal antibody. Not only are these used now for many diagnostic techniques, they are also being used to very specifically kill cancer cells. (The new pharmaceutical drugs with names ending in –ab are monoclonal antibodies.)

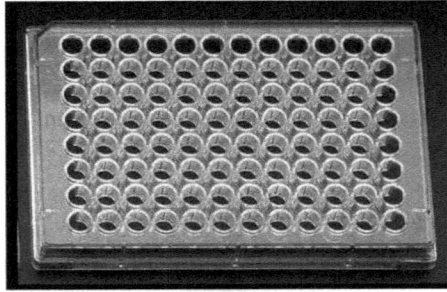

Figure 6.2. A multiwall plate. A different sample can be placed in each well, and a multi-tip pipetter can be used to fill each well with the same solution. Often, a color is developed from a specific reaction, and the color in each well can be read by an automated device, greatly reducing the manual labor involved.

However, one last question remains, how to use the antibody to identify particular proteins within cells or tissues. The monoclonal antibody is, as you might guess, quite expensive. Since most of them are made by immunizing mice against materials to be tested, and mice are quite small, requiring small amounts of original material or antigens, monoclonal antibodies are also in short supply. However, antibodies are proteins, and antibodies recognize (normally foreign) proteins. It is therefore possible, for instance, to immunize a horse against a mouse antibody.

At this point it is useful to use another term for the mouse antibody to emphasize that the antibody is itself a protein. Antibodies are a type of protein called immunoglobulin, and one can immunize a horse against mouse immunoglobulin and collect horse anti-mouse antibodies. Horses are big animals, and it is possible to collect large amounts of horse anti-mouse antibodies, enough to attach to the horse antibody a label by which we can recognize it. Such labels can be enzymes that can carry out color-producing reactions, radioisotopes, or fluorescent molecules. Companies now make and sell commercially both the mouse monoclonal antibodies against specific proteins ("primary antibody") and the labeled large animal anti-mouse antibodies ("secondary antibody"). Now, to detect the first appearance of a molecule in a specific cell or tissue of an embryo, or a protein that characterizes and can localize a cancerous or otherwise diseased cell, one undertakes the following multi-step procedure:

- Add the specific mouse antibody to the tissue to be analyzed and allow it to interact
- Wash well to remove any unbound antibody
- Add the labeled horse or large animal anti-mouse antibody and allow it to bind to the mouse immunoglobulin
- Wash well to remove any unbound antibody

- Conduct any further steps (expose to x-ray film, carry out enzymatic reaction, observe under UV light to cause fluor to fluoresce) to reveal the label.

Of course the entire procedure must be done with considerable caution and control procedures to rule out contamination, false positives, false negatives because of too vigorous washing or other problems, and other sources of error but, if done correctly, it is marvelous. One can detect, for instance, a change that takes place over a few minutes in two or three cells such that a cell in an embryo suddenly decides to become a muscle or a nerve—and then go on to the next question of what makes it decide to do that. To detect differences between two cells over a very brief interval: We're not in Kansas anymore.

Recombinant DNA

A second fantastic technique was the ability to control genes and to identify when a gene was active or not. The whole process of identifying genes, isolating them, reading their sequences, cutting them up, and recombining them is a very long and exciting saga, worthy of discussion later (pp. 80-119). In the current context, how to create exquisite sensitivity, we are interested in the last point, recombination of genes. For a developmental biologist, it is important to know when and where the first proteins, for instance to make a muscle or nerve, are formed, but it is equally important to know when and where the genes that produce these proteins are activated or deactivated. For, after all, with very few exceptions all cells in an organism contain the same chromosomes and the same genes, but the difference between a muscle, nerve, skin, liver, testis, ovary, or any tissue is which genes were active to form the tissue or organ and which genes remain active, and which genes are silent. It is the whole point behind the research goal of finding stem cells, those cells that would have the ability to regenerate any organ. We can of course look for the product of gene activity, messenger RNA. We can do this by exploiting the fact that messenger RNA (mRNA) is a variant copy of the DNA that produced it and will bind to the DNA in the same manner that one strand of DNA binds to another to form the famed double helix. We can make a single-stranded copy of the original DNA and allow it to bind to any RNA that is present in the cell. This works, but it requires high temperatures, which may have other effects, and for many purposes is not very sensitive.

A far cooler approach is to make the DNA show itself when it is active. This depends on a peculiar property of our DNA. Most of it does not "code" or carry instructions for anything. The DNA that actually carries the instructions for making proteins in our body is less than 2% of our total DNA. The rest of it used to be called "junk DNA" since we had no idea why it was there. We now know that a large part of it is used to tell the coding section—the gene that actually produces something—when to be active. If we picture a gene as reading from left to right, a very large part of the DNA to the left of the gene, sometimes some of the DNA in the middle, and sometimes some of the DNA to the right are responsible for telling the gene when to be active. The part to the left is usually called the promoter region and it has sequences of DNA that can bind many things. For instance, in cells that respond to steroid hormones (sex and adrenal hormones), the hormones when present bind to proteins that, carrying the steroid, can bind to specific promoter regions called steroid receptor regions. Other regions can respond to different amounts of calcium inside the cell, metabolites that signal how much nutrition is available, and many other molecules that define in which cell the DNA resides and under what circumstances. When, in developing embryos, some cells "commit" to a particular path---that is, that they will become insulin-secreting cells and nothing else—large sections of the promoter DNA that in this situation are not needed are effectively buried in tightly-binding proteins and cannot easily be unburied. So, now that we can cut DNA at specific places and splice it back together (see Mr. Potato Head, p. 121) it becomes possible to add to a cell a second copy of a promoter sequence to which is attached a molecule that serves as a label. When the gene of interest is active, the second promoter sequence will produce a detectable "label" protein that will mark the fact that the gene is active.

Genes and promoters

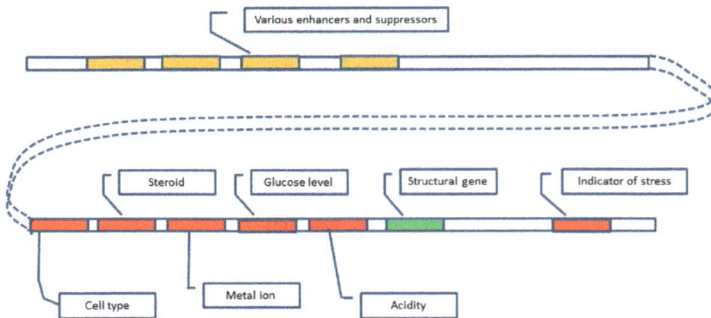

Figure 6.3. Structural genes (green, the genes that produce specific proteins) are only a small part of the gene complex. In the conventional manner in which DNA is diagrammed, to the left (upstream) of the structural gene are a variable number of promoters (red), each of which responds to a specific signal, and contributes to the activation or suppression of the structural gene. Less commonly, promoters may be found within the structural gene or, as illustrated, to its right (downstream). Elsewhere on the chromosome may be enhancers or suppressors, which adjust the level of activity of the structural gene. In general, the chromosome is folded in such a way that the enhancers and suppressors affect the ability of other molecules to reach the structural gene and its promoters.

Fluors

A favorite label at present is the protein that makes jellyfish glow. This can be attached to a promoter (see Chapter 9, p. 114) for a description as to how this is done) in which one is interested and inserted into a cell, animal, or plant. The cell now contains two promoters. One turns the gene on when it is supposed to be turned on, and the cell therefore behaves normally. The other turns on the synthesis of the jellyfish glowing protein. If we illuminate the cell or the organism with ultraviolet light, if the gene is on the cell or even organism glows. The original version, green fluorescent protein, was originally inserted in this manner into a tiny worm by Martin Chalfie, but now there are many variants and colors of fluorescent molecules, and they can be used for many purposes, including being inserted into animals[34] or cells to identify when, where, and in what circumstances, certain genes are activated in even single cells. Again, compared to the science of the 1960's, we are not in Kansas anymore and we can read details at a sensitivity a billion times more than half a century ago.

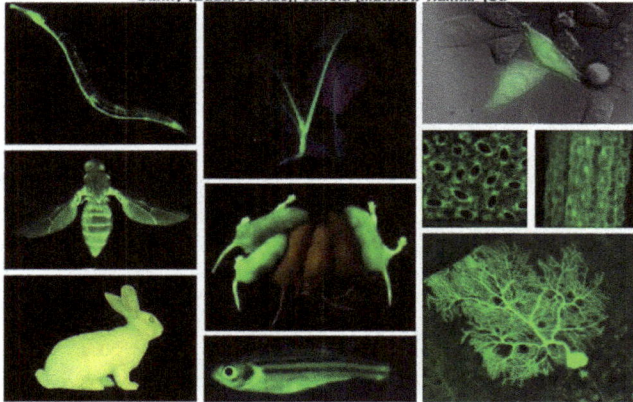

A gallery of GFP images. {Photo credits by columns left to right: C. elegans (John Kratz, Columbia University), Drosophila (Ansgar Klebes, Freie Universitaet, Berlin), Alba the GFP bunny (Eduardo Kac), canola [Matthew Halfhill (St.

Chalfie M PNAS 2009;106:10073-10080

PNAS

Fig. 6.4 A gallery of GFP images. {Photo credits by columns left to right: C. elegans (John Kratz, Columbia University), Drosophila (Ansgar Klebes, Freie Universitaet, Berlin), Alba the GFP bunny (Eduardo Kac), canola [Matthew Halfhill (St. Ambrose University, Davenport, IA) and Harold Richards, Reginald Millwood, and Charles Stewart, Jr. (University of Tennessee, Nashville)], mice (Ralph Brinster, University of Pennsylvania, Philadelphia), zebrafish (Brant Weinstein, National Institutes of Health, Bethesda), cultured HeLa cells (Jerry Kaplan and Michael Vaughn, University of Utah, Salt Lake City), Drosophila embryonic cells (Jennifer Lippincott-Schwartz, National Institutes of Health), Arabidopsis thaliana hypocotyl cells (David Ehrhardt, Carnegie Institution of Washington, Stanford, CA), and mouse Purkinje cell (National Center for Microscopy and Imaging Research, University of California, San Diego).} From Proc. Natl. Acad. Sci. vol. 106 no. 25 Martin Chalfie, 10073–10080 (2009)
http://www.pnas.org/content/106/25/10073/F10.expansion.html

Chapter 7: Mechanisms of protein synthesis

Ribosomes, mRNA, tRNA

"The genetic code is universal." What this means is that in all organisms, the sequence of amino acids that will make a protein is encoded in the DNA, with a sequence of (mostly) non-overlapping three bases in DNA representing a single amino acid. Thus, to make a three-amino acid peptide such as phenylalanine, methionine, serine, the DNA would read adenine, adenine, adenine; thymidine, adenine, cytidine; thymidine, cytidine, guanidine. The universality is that this sequence would be valid for bacteria, corn, fruit flies, and humans. To understand how astonishing it is to find that the genetic code is universal, we need to make a brief excursion into the molecular biology of how proteins are synthesized. Though when understood it makes sense, other mechanisms are possible, and the common mechanism was not predicted before it was worked out.

The genetic information is in the DNA in the nucleus of the cell. DNA cannot leave the nucleus, but proteins are synthesized in the cytoplasm outside of the nucleus. To make a protein, therefore, the cell has to make a copy of the DNA that can be exported into the cytoplasm. This copy is made, similar to the second or noncoding strand of the DNA, in a different kind of nucleic acid called RNA. There are several types of RNA. This one, which carries the information from the DNA, is called messenger RNA or mRNA. It is a transcription of the DNA, in the sense that, if you were to copy directions given in a language you didn't understand, you would write down what you heard (transcribe what you heard) in the hopes that someone, recognizing the sounds, would be able to translate it for you later. The translation from nucleic acid language to protein language takes place in the cytoplasm.

> In France, you ask directions to the train station and hear the reply, which you TRANSCRIBE as "Vou zalley too dra, trwa cent metres, et tourney za gosh. La port ey devan vou."

> You have no idea what this means, but your friend speaks French. She looks at it, recognizes your transcription as "Vous allez tout droit, 300 mètres, et tournez à gauche. La porte est devant vous."

> She TRANSLATES to you, "You go straight ahead, 1000 feet, and turn left. The door is in front of you."

The mRNA can be a very large molecule, and the machinery to manufacture the protein can be complex. The mRNA is therefore carried to a very large (in molecular terms) machine called the ribosome. The ribosome acts like a big press in a factory: given material supplied to it, it stamps out parts. The parts that are made depend on the template provided to the machine. Here the mRNA is the template. Since the machine is not readily moveable, the raw materials are brought to the machine. In the cell, the role of the stockboy who delivers the raw materials is played by another type of RNA called transfer RNA (tRNA). With the mRNA sitting in the ribosome, the tRNA trundles the appropriate amino acid to the ribosome, and each amino acid is attached to the end following the sequence dictated by the information on the mRNA. This is the translation: nucleic acid language is translated into protein language, and the protein is formed.[35]

The tRNA is quite a remarkable molecule. It is the translator. There are enough different tRNAs to account for all natural amino acids. Each tRNA can hold a single amino acid at one end, while at the other end of this molecule is a sequence of three bases that can pair with three bases in the mRNA, and hence derived originally from the sequence in the DNA. This sequence of three bases on the end of one tRNA is unique for each type of tRNA, and is specific for the type of amino acid that it carries. This is the genetic code: a specific sequence of three bases, for instance adenine, uridine, guanidine or AUG (in the mRNA, hence TAC in the DNA and UAC in the tRNA) represents the amino acid methionine and only that amino acid. Wherever the sequence AUG appears in the mRNA, methionine is inserted into the growing protein chain.

There are several aspects that make this a remarkable story. The first is that I did not make any mention of any specific animal, plant, or bacterium. That is because the code is universal. AUG represents methionine in a bacterium, in a tomato, in a fruit fly, or in a human. This is the true sense of the analogy that it is more remarkable than if a Martian came to Earth and proved to natively speak idiomatic 21st C New England American English. There is no reason that it should not speak German, Chinese, Xhosa, or a non-human language. We can design tRNAs with the wrong base sequence, and they work well in artificial systems, inserting the wrong amino acids at the appropriate places (for obvious reasons, in living organisms they are lethal) demonstrating that there is no physical reason why the base sequence to amino acid pairing is obligatory, but it is. It is the single strongest argument that we have that, if life arose more than once on earth, at least everything we see today on earth is a descendent of only one original form of life and has evolved from that original form.

The genetic code

The most striking advances of the late 20th C included the rise of molecular biology. This is a fantastic and rich story. The story of how this was worked out is fascinating in its own right and is described in Chapter 8, "The Rise of Molecular Biology" (p. 89), specifically starting with Cool Trick #5, p. 101. But the importance to the story of evolution is special. As far as we can tell, there is no reason why there cannot be alternative means of making proteins, and no physical reason why a tRNA that has a specific three-base sequence at one end must carry a specific amino acid at the other end. But all living things use the same basic machinery, and all translate the code in the same fashion. With very few and minor variations, the tRNA from a bacterium can be used to synthesize protein using mammalian machinery and mammalian mRNA. Unlike the situation for water and carbon[7], which we believe to interact in such a specific fashion that we consider that any life in the universe will use water and carbon, the coding seems to be arbitrary. And yet it is universal. This is one of the strongest possible arguments for the relatedness (and common descent) of all living organisms.

Gene composition and arrangement

If the universality of the genetic code were not astonishing enough, our recognition of how the DNA is constructed provides the final, powerfully convincing, argument. In the 1970's several Drosophila geneticists, including Walter Gehring in Switzerland, identified and sequenced the first genes known to control patterns, such as why a fruitfly starts at one end with a head and antenna, continues with a thorax, and ends at the other with an abdomen. They called these genes homeotic genes, referring to the fact that mutations of these genes could create monsters with feet growing where their antennae should be, or an extra segment of thorax with an extra set of wings. Within each of these genes was a short run of bases that encoded portions of proteins that allowed the proteins to bind to DNA. The proteins produced by these homeotic genes would bind to promoter regions of groups of other genes, activating or deactivating them and causing cells to follow one developmental pathway as opposed to another. The fact that all of the homeotic genes had similar short sequences was interesting in its own right, but it was certainly intriguing that these several genes were lined up on a Drosophila chromosome in a linear order that reflected the head-to-tail structure of the animal itself.

What was even more astonishing was that, once the sequence of homeotic genes was identified, the genes were sought and found in other animals. Even in humans, the sequence was a bit more complex but was found, and, as in fruit flies, the sequence represented genes that controlled the organization of the body, from head to rump. These were what we call conserved genes—genes whose function is so critical to the well-being of the animal that almost any mutation is lethal, and so they are passed along almost intact throughout our entire evolutionary history. Think of that: genes very similar to those that tell a fruit fly embryo to plan for antennae in one region are those that determine where our heads will be. This is true for many other genes as well. A gene that is important in the formation of our eyes, such that if it is mutated people have small to non-functional eyes, is extremely similar to the gene that puts eyes in their proper place and form in fruit flies. This is true even though the eyes are extremely different in construction. The similarity is so strong that the normal human gene can be put into a fly carrying a mutated form of the fly eye gene, and the human gene can come at least close to helping the fly produce a normal eye. We can also identify many genes that are highly conserved from animal to animal. By comparing their base sequences and where they sit in the chromosomes relative to other genes, we can establish a line of descent (the most similar forms are the most closely related, *etc.*) that either corresponds exactly to the anatomical and fossil records or, where it differs, indicates that our interpretation of these latter two must be adjusted. It is virtually impossible to imagine a scenario of creation that would produce these similarities without invoking evolution. Otherwise, why should God use the same genetic tools to design the faceted compound eye of a fly, which gathers images more or less like a light cable, and our eye, which focuses images onto a single retina? It boggles the mind.

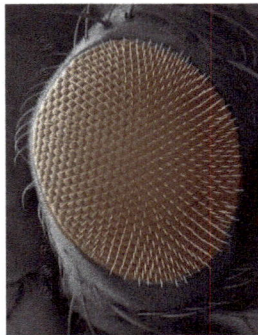

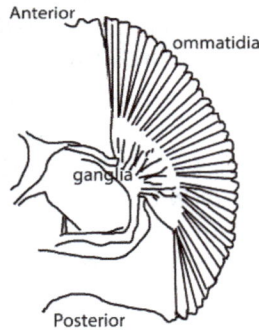

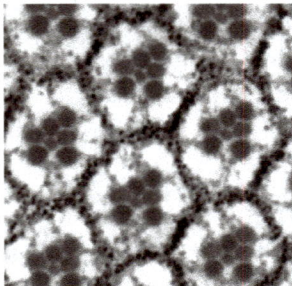

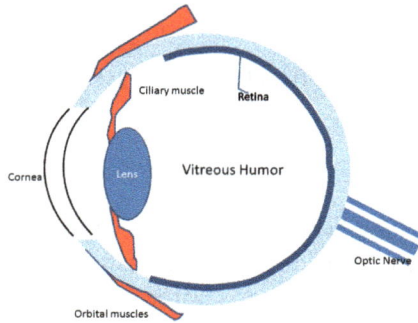

Figure 7.1. Upper row: The eye of an insect consists of a collection of ommatidia, or individual lens units, each of which produces a small segment of the image, which is passed in packets through different layers of nerves in the ganglia. An eye is seen in cross-section on the right. Middle row: Each ommatidium consists of eight cells plus associated pigment (to isolate and sharpen the image), seven of which surround the center cell, which is important in development because it initiates the development of the ommatidium. Bottom: A mammalian eye, in contrast, consists of one lens, which can be adjusted and moved to change the point of focus, and which projects to a single retina and, following some selection and interpretation, thence through the optic nerve to the brain.

~~~~~

Chapter 8: The Rise of Molecular Biology

A question of aesthetics

The rise of molecular biology, which validated beyond doubt the hypothesis of descent with modification, coincided with a transition in American biomedical science. Coming from a European- style aristocratic, patrician, contemplative style, in which experiments were cogitated and planned with care, with few experiments undertaken, it became a more peasant-style, "suck it and see," "Gee, we can play in this mud pile and get paid to do it?" approach driven by a rising and increasingly influential immigrant generation. Such a change in approach was inordinately successful. Perhaps newer immigrant and rising generations will likewise impose their styles and build new sciences. In any event, the new science was built on remarkable insights and innovations, sometimes as simple and as elegant as kitchen chemistry. The story is a marvelous tribute to the importance of playing in building science. One could even describe the rise of modern molecular biology as a series of cool tricks developed by those class clowns who were always pulling off some elaborate mystification of their friends. To understand this story it helps to realize that science is an onion. The questions have existed since the dawn of history. At the first instance, the questions are there, but there are no techniques available to answer the questions. When the techniques are developed, we peel the first layer of the questions, only to find another layer of questions underneath. Those questions provoke new techniques, which yield new answers, which yield new questions. It is not uncommon for a scientist to pursue the same question throughout his or her life. Many a scientist has been asked (I have), "Haven't you answered that question yet?" To give an illustration of the onion-like quality of science: A child may ask, "Why are rabbits brown?" A parent may respond, "Because God made them brown." To some, this may be a satisfactory answer. Others may pursue, "But why—questions beginning with 'why' are typically poor scientific questions, but we will tolerate it here—did God make it brown?" One could turn to analysis of predator-prey relationships, protective coloration, and mechanisms by which natural selection will favor the best-camouflaged rabbits. Alternatively, on a more mechanistic basis, one could pursue the genetics of brown pigment, the developmental biology that determines where on the body the pigment will be distributed and when and how intensely it will appear. Geneticists and developmental biologists will wonder why the pigment appears only in the skin, not in internal organs. One might ask how brown pigments are synthesized, what enzymes are needed, what colorless substrates are converted into colored pigments, and how the pigments are distributed (contained in little granules) within cells. The chemist will ask why it is that some substances are colored and others are not. The physicist will try to understand why certain atomic configurations within molecules absorb certain wavelengths of light and reflect others, producing color, and the physiologist will be concerned whether the absorption of light produces any biological effect, ranging from mood swings to cancer. At each level some people will say, "above this line is very interesting; below this line is too dry/abstract/mathematical for me." All the questions are valid, and it is the aesthetic decision of where to draw the line that distinguishes poets, philosophers, theologians, evolutionists, geneticists, developmental biologists, biochemists, chemists, and physicists. In an analogous manner, the recognition that, in the Mendelian sense, genes exist, raises new questions: what the genes are, what their structures are, and how they work. Attacking these questions, and peeling them back, layer by layer, became the work of molecular biology. The first questions became to determine the physical nature of genes, their location in the nucleus of cells, and then to determine how they worked.

8.1. Giant chromosomes from a fruit fly. Fruit flies, for unknown reasons, have giant cells in their glands, and in these cells the DNA increases but stays together, so that it builds giant chromosomes of over 1000 strands each.

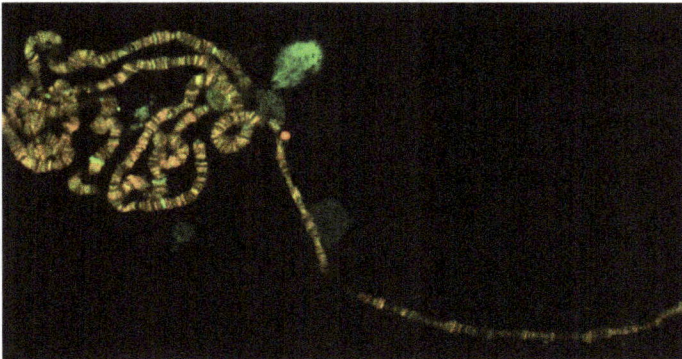

In this picture, in which the chromosomes are stained with fluorescent markers, the variegation indicates substructure. In fact, each patch represents only a few genes, which are arranged in linear order. Variations in the pattern were extremely helpful in understanding how genes were arranged on the chromosomes.[36]

The first "really cool trick" was to determine the chemical nature of genes. Because of a peculiarity of flies such as the fruit fly *Drosophila*, the salivary glands of which have giant cells containing giant chromosomes (Fig. 8.1), T. H. Morgan and others had determined that genes existed on chromosomes and that they were arranged linearly along the chromosomes. The simplest cells in which we find chromosomes are sperm, which are essentially cells containing packed chromosomes, a motor to drive them toward the egg, and a penetrating device so that they can enter the egg. One can describe a sperm as a bunker-busting missile: it has a penetrating device (enzymes to digest its way into the egg, called the acrosome), a warhead (the packed chromosomes containing its DNA) fuel (provided by a large mitochondrion) and a rocket motor (the sperm tail), and that is all. But we have over 20,000 genes, and sperm don't have enough behavior to allow us to deduce what the genes do. Following the development of an animal is far too complex to get a direct answer. There had to be a simpler way. Sperm, and chromosomes, are made of a mixture of proteins and deoxyribonucleic acid (DNA). Chemists could already say a lot about what these molecules are. They are macromolecules, giant arrays of thousands of individual atoms. Molecules are tightly-linked arrays of a few atoms. For instance, common table sugars consist of twelve carbon atoms, twenty-four hydrogen atoms, and twelve oxygen atoms. Propane gas consists of three carbon atoms and eight hydrogen atoms. Octane, a constituent of gasoline, consists of eight carbon atoms and eighteen hydrogen atoms. It is less volatile than propane because it is bigger and heavier. Macromolecules are often arrays of large numbers of molecules. Proteins, for instance, consist of basically linear arrays of molecules called amino acids:

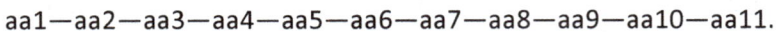

$$aa1-aa2-aa3-aa4-aa5-aa6-aa7-aa8-aa9-aa10-aa11.$$

DNA consists of a linear array of a sugar called deoxyribose, linked by phosphates. Each sugar has also attached, hanging on to its side, another molecule called a base:

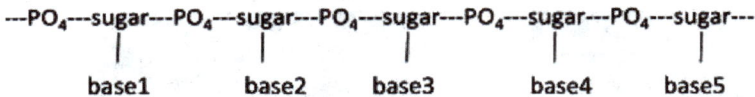

$$---PO_4---sugar---PO_4---sugar---PO_4---sugar---PO_4---sugar---PO_4---sugar---$$

| base1 | base2 | base3 | base4 | base5 |

Figure 8.2. A schematic diagram of the structure of DNA and RNA. Sugar phosphates are linked head-to-tail. The sugar is ribose for RNA or deoxyribose for DNA. To each sugar is attached a "base," a one- or two-ringed molecule called a nucleotide. There are 4 different types of nucleotides that may be attached to the sugars: adenine, guanidine, thymidine, and cytidine for DNA. In RNA, uridine replaces thymidine.

DNA as the genetic material

Prior to the Second World War, to most scientists it was pretty clear that the genes had to be in the protein rather than the DNA, for the simple reason that DNA was boring. There were 20 different types of amino acids in protein, and only 4 different types of bases in DNA. Simply put, you can't make a language with only four letters. Even if you give yourself two consonants and two vowels, even if you reuse letters your range is limited: abce could give bace, baec, caeb, ceab, ebac, ecab, baab, baac, etc. With twenty letters, you can build a language. (The last sentence contains 16 letters.) How could you carry all the information necessary to construct an organism if you had to write that information in a language that consisted of only four letters? However, by the 1940's evidence was beginning to accumulate that, disturbingly, suggested the opposite. The first to argue that DNA, not protein, was the genetic material, can be considered to be the first really cool trick:

Really cool trick #1: The evil scientists create a bad bug.

Avery, MacLeod, and McCarty[37], working at what is now Rockefeller University, were interested in why some forms of pneumonia would kill and others would not. They could produce two types of pneumonia in mice. The first, caused by a bacterium that grew in a petri dish in smooth, round droplets, invariably killed the mice. However, a variant, mutant form of the bacteria, which grew in rough-edged, less symmetric colonies, caused only a mild infection from which the mice recovered. Obviously there was value in knowing how they differed. Killing the bacteria by boiling them obviously prevented the bacteria from infecting the mice. Injecting dead bacteria into mice produced no effect. However, the experimenters wanted to know if the dangerous, "smooth" bacteria produced a toxic chemical that made the infection more virulent. They therefore injected into mice a combination of killed smooth bacteria and live "rough," non-virulent bacteria. In this situation the mice developed a severe pneumonia and died. However, there was a surprise involved. When they took bacteria from the dying mice and injected these bacteria into new mice, the second-round mice also died. If the smooth bacteria had merely delivered a chemical to assist the rough bacteria, the chemical should have been diluted in the first mice and ineffective in the second mice. Avery and MacLeod therefore grew the bacteria from the second-round mice in a Petri dish. These bacteria grew in the form of the smooth bacteria and in every respect behaved like the virulent strain. But the virulent strain that had been used had been boiled and was dead. Something had been transferred from the dead bacteria to the live ones and had transformed the rough, non-virulent strain into the smooth, virulent form. Since these bacteria continued to grow as smooth, virulent bacteria, by definition what had been transferred was genetic material or genes. It would be possible to determine chemically what this material was.

A brief note of explanation: The smooth form owes its appearance as a colony to a gelatinous coating that creates the smoothness and also prevents the immune system of the mouse from destroying it. Thus the gelatinous coat is also the source of its virulence. The rough form has a mutation that prevents it from making the gelatinous coat. Without the coating, the macrophages (bacteria-eating cells) of the mouse can destroy it and contain the infection. The transformation consisted of the restoration in the rough cells of the ability to make the gelatinous coating.

They prepared extracts from the boiled smooth bacteria that contained either DNA or protein and injected these extracts, rather than the dead bacteria, along with rough bacteria into mice. The extracts that contained DNA transformed the rough bacteria into the lethal smooth form, but the extracts containing protein did not.

Still, the logic was unacceptable. Most scientists concluded that the techniques were limiting: it was not possible to remove the last few percent of protein from the DNA extract. Surely the genetic material was that last bit of protein, a sort of super-protein, that for some reason clung very tightly to the boring DNA. The experiment had demonstrated that inheritance, at least in bacteria, was handled not by a mysterious vital factor, but by a chemical. Which chemical it was remained an issue of some dispute. It was not until 1952 when Alfred Hershey and Margaret Chase used a new technology to provide more convincing evidence.

Really cool trick #2: Bacterial milkshakes

Hershey and Chase took advantage of a newly developing technology and a biological peculiarity. The atomic era brought not only bombs but reactors for power generation, and the reactors produced waste materials that included different types of radioactive atoms. Some of these could be used by researchers. Radioactive atoms are usually processed in living organisms exactly like their normal, non-radioactive, brothers, but by use of appropriate detectors they can be located within the animal, its cells, or extracts of chemicals from its body. Using radioactive elements gives researchers two huge advantages. First, it is possible to know, for instance, that a particular protein was formed at least in part from a sugar molecule that had been fed to the animal the previous day. It is possible to trace the synthesis and degradation of molecules of biological interest. Second, the measurement of radioactivity is quite sensitive, improving our ability to measure trace amounts or contaminants up to 10,000 times the sensitivity that we have for purely chemical means. The biological peculiarity was that even bacteria have parasites that feed upon them. Some of these parasites are viruses called bacteriophages (literally, bacteria-eaters). One of these bacteriophages (phage for short) is a lollipop-shaped virus that attacks a very common gut bacterium. When it attacks the bacterium, the lollipop attaches to the outside of the bacterium, stick end first (Fig. 8.2). Then it injects something into the bacterium through the stick, which is really more like a straw. Whatever gets into the bacterium immediately sets about to produce more phage, until the bacterium fills with new phage and bursts, releasing the new phage into the medium to find more bacteria. The whole process takes about 20 minutes, and whatever gets into the bacteria carries, by definition, the genetic information to make more phage.

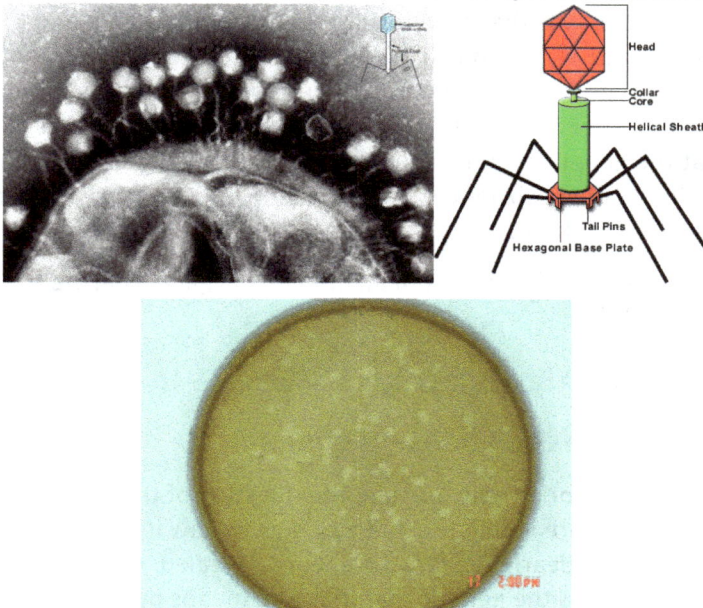

Figure 8.3. Bacteriophage. Upper Left: Bacteriophage attached to the cell membrane and cell wall of a bacterium, as seen in an electron microscope. Upper Right: The general structure of a bacteriophage. Lower: Measuring bacteriophage. In this experiment, viruses were scattered on a "lawn" of bacteria, otherwise described as an even coating of bacteria growing on medium in a Petri dish. The bacteria are stained and look dark in the picture. Where a virus has landed, it has infected the bacteria, grown, and reproduced, killing the host and infecting the bacteria next to it. Thus small circles of killing appear, as is marked by the clear areas, or plaques. Each plaque represents the descendants of one virus, or a clone.

The question was, what was injected? Chemically, DNA contains a lot of phosphorus, which is found only sparingly in a few proteins. Proteins, on the other hand, contain some sulfur, which is not found in DNA. Happily, there are radioactive forms of both phosphorus and sulfur. What Hershey and Chase did was to grow bacteriophage in the presence of radioactive phosphorus and sulfur until it was certain that the phage had plenty of both types of radioactive elements. They then used these radioactive phage to infect bacteria. They allowed the infection to start but, before the phage could reproduce and kill the bacteria, they made a milkshake. Literally, they threw the culture into a common kitchen blender and ran the blender. This shreds the bacteria as it does fruit or vegetables. Next they put the shredded mix—we call it a homogenate—into a centrifuge and centrifuged it. The bacterial cell walls, to which the lollipops were attached, were pushed to the bottom of the centrifuge tubes, while the insides of the bacteria, which are more liquid, stayed suspended in the fluid, or supernatant. Hershey and Chase then looked to see where the phosphorus and sulfur were. Their results were unequivocal. All the sulfur, thus protein, was spun out with the cell walls and therefore had not entered the bacterial cells, while the phosphorus, thus DNA, was found with the material from inside the cell, suspended in the supernatant. The DNA was what had entered the cell and therefore had to be the genetic material. This was the same result that Avery, MacLeod, and McCarty had found, but the sensitivity was much higher. Instead of a possible contamination of a few percent, it was now clear that the amount of phage protein inside the cell was less than tenths of a percent. The evidence was in. One had to accept the evidence that the genetic material was DNA, and to understand how this seemingly uninteresting substance could carry the information to make a phage, a bacterium, and even a human.

The next big breakthrough came from an intellectual game that we can call "molecular billiards".

Really cool trick #3: Molecular billiards—the logic supporting DNA as the basis of genes

Many research laboratories now pursued the question of the structure of DNA, hoping that the structure would provide a clue as to how it could encode genes. The surprise was how clearly the structure, once revealed, supported the argument that DNA was the genetic material. The means by which the structure was revealed could be called molecular billiards. Molecular billiards is a means of interpreting molecular structure. It depends on bouncing X-rays off of atoms. How is this done? The trick is the following: X-rays have wavelengths close to the size of the spacing of atoms in a molecule. Waves this size tend to ricochet off molecules, careening off at an angle to their original trajectory. Waves also have the property that when they collide with each other the result is a wave that is the sum of the original waves. In other words, when two waves meet when each is at its peak, the resulting wave has a peak that is twice as high; when both are at their valleys, the valley is twice as deep. When a peak meets a trough, the two waves cancel out. You can see this effect by throwing two stones simultaneously into a pond and watching as the radiating waves meet, or by watching two waves encounter each other in the ocean (Fig. 8.4).

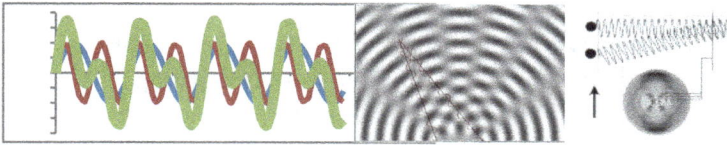

Figure 8.4 Wave interference. Left. When two waves of different frequencies (red and blue) meet, when they are in phase, they reinforce, increasing the strength of the positive or negative phase (green). When one is up and the other is down, they cancel each other (green line near the zero-level horizontal line). Middle: Demonstration of this effect using two sources of vibrating rods to generate the waves. The two rods are at the bottom. Right: how X-ray crystallograms are produced. X-rays are shot toward the crystal as if the X-ray source is situated in front of your nose as you read this and the crystal is between you and the paper or screen. They are deflected by successive rungs of a molecular helix, and they travel, slightly out of phase, toward an X-ray film behind the paper or screen. If they meet the film when they are in phase, they expose the film and create a dark spot. From the position of the spot and its distance from the center, it is possible to calculate the dimensions of the repeating unit that produced it.

Noise cancelling headphones work on the same principle. By generating sound waves out of phase (i.e., up when the incoming wave is down; down when the incoming wave is up) they produce a wave of net zero amplitude, or silence. Biophysicists had exploited this property by making crystals. In a perfect crystal, atoms or molecules are aligned in an exact order, providing enough atoms to be able to detect the effect. Linus Pauling had made crystals of proteins and, using the properties of waves, deduced the structure. Using the summing properties of waves, he had recognized that the spots produced on an X-ray crystallogram (the photograph produced) represented repeating units in the proteins, specifically, the successive loops in a helix, and he had recognized one of the major configurations of proteins, called the alpha-helix. (He had originally identified several possible types of repeats that could produce such reflections, named successively by Greek letters. The first one proved valid.) To make a crystallogram of DNA required a good crystal, a very difficult task to achieve. Finally, Maurice Wilkins and Rosalind Franklin produced crystals good enough to submit to X-rays, and they produced a usable crystallogram. The crystallogram showed the characteristic repeat distances, indicating that DNA existed in helical form. The question was, what kind of helix?

The first clue came from the density of the crystal. Helices have the property of being able to mingle, that is, to slip in among each other. Picture ten springs in a box. Each of the springs is separate, and each spring weighs 50 grams. Suppose that the box is filled, and that it is 10 cm x 10 cm x 10 cm (about 4.5 inches on a side). The density, not counting the weight of the box, is 500 gm/1000 cm³ or 0.5 gm/cm³. Now suppose that the springs are packed so that they make a group of 10 springs, only slightly larger than one spring by itself. They still weigh 500 gm, but their volume is now closer to 100 cm³, and the density is now 5 gm/cm³ (Fig. 8.5). Thus, if you can get a good, clean crystal, and you know how much one strand of DNA should weigh (calculated by summing the weights of the atoms in it), the density of the crystal should tell you how many strands make up the helix. When this measurement is done, it turns out that DNA should consist of a two- or three-stranded helix, definitely not one or four strands. Starting with this information, and the information read from the Wilkins' and Franklin's X-ray crystallograms concerning the distances between repeats of the helix, James D. Watson and Francis H. Crick literally began to assemble models of DNA strands to see if they could replicate the characteristics and generate a model of DNA. Most of the models did not work. One of the models, for a two-stranded structure of DNA, did have the right spacing, but it also had further constraints. Because of the way that atoms link together, the side chains or bases had to point inward, or into the core of the helix. One of the constraints was size. The bases come in two sizes. Two of the four bases, adenosine (A) and guanosine (G), are big and the other two, cytosine (C) and thymosine (T), are smaller. There is not enough room down the core of the helix for the two big ones to fit across from each other. The two smaller ones can fit across from each other, as can one big and one small, but there is a further constraint. Molecules can have local charges, somewhat like magnets, though in molecules they are more like the positive and negative poles of batteries. Like magnets, if you have two positive charges next to each other, they will push each other away; the same applies to two negative charges. The two smaller bases run into this problem, as do some of the other combinations. In fact, there are only two ways that the helix will work: if an adenosine is across from a thymosine, or if a guanosine is across from a cytosine. All other combinations fail either because of size or because of where the charges are (Fig. 8.6).

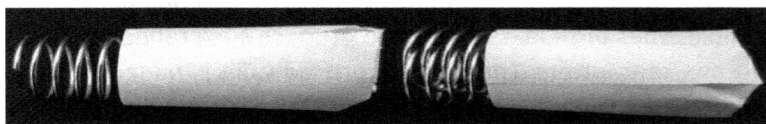

Figure 8.5. Two springs (helices) can be wrapped into approximately the same volume that is occupied by one spring. Since the density of the object (a crystal of DNA, represented by the paper sheath) is the mass (weight) of the object divided by its volume, the density of two-spring package will be nearly twice that of that of the one-spring package. From such measurements it was possible to calculate that DNA was configured in a helix that consisted of more than one but fewer than four chains. Another trick would be to suspend the crystal in liquids in which it will not dissolve, but of different densities. If it is less dense than the liquid, it will float, and if it is more dense, it will sink. This is the same type of analysis that Archimedes used to determine the amount of gold in the king's crown. (He jumped out of his bath and went running to tell the king, shouting "I found it!" [Eureka! In Greek].)

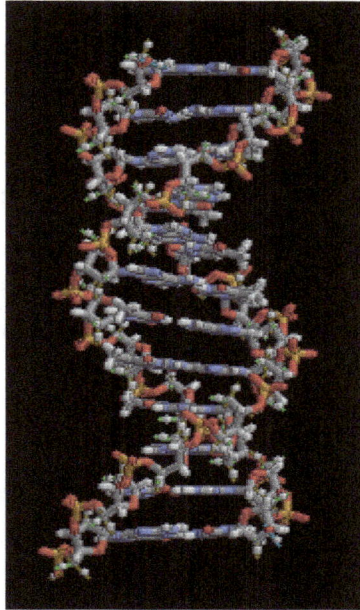

Figure. 8.6. A representation of a DNA molecule. The sugar chains are in red and the paired bases between the chains are in blue. For the rotating image, which gives a good impression of the spatial structure of DNA, see the hyperlink, http://en.wikipedia.org/wiki/DNA.[38]

Watson and Crick saw this restriction and realized that it explained a curiosity—recognized as potentially meaningful—about DNA, called Chargaff's Rule: that, in DNAs taken from different organisms, the absolute proportions of the bases could vary, but the amount of A always equaled the amount of T and the amount of C always equaled the amount of G. More importantly, they realized that this relationship conferred an extremely important property to the DNA. If you could pull the two strands apart and reassemble new strands on the newly naked single strands, the only new strands that would work would have to be identical to those that had just been pulled off.[39]

In other words, each strand was a template of the other, and the new strands would have to be replicates of the old strands. Put even another way, you didn't need an instruction sheet to make the new strand. If you had even a nonspecific zippering mechanism, you could make a new strand with a predictable sequence of bases. In other words, the molecule could be replicated without further instruction. This was a huge insight, and it made it intellectually necessary for DNA to be the genetic material, because this mechanism resolved what was called the Russian Doll problem. You know the Russian dolls, or matrioshkas: one doll nested inside another (Fig. 8.7). To biologists, it meant the following: if the genetic information was carried on protein, what protein carried the information to make the protein that carried the information? And what protein carried the information to make the protein that carried the information to make the protein that carried the information? And…. In other words, there had to be something fundamental that did not require other information to build it. And for proteins, there was no explanation. The resolved structure of DNA gave such a mechanism: one could build a new strand of DNA by following the template of the old strand of DNA. The molecule could replicate itself. This property by itself made it possible for DNA to be the genetic material. Watson and Crick appreciated this interpretation, as they expressed in a memorable sentence, "It has not escaped our attention that the specific pairing we have suggested immediately suggests a possible copying mechanism for the genetic material."[40] This was the secret of the double helix, usually missed by people who quote or mention it. No other model got out of the Russian doll problem, and this solution was particularly elegant. The next steps were to prove that the model was valid, and then to figure out how it worked.

Figure 8.7. The Russian Doll Problem. The philosophical question was, "What carries the information to make the molecule that carries the information to make the organism?" The facts that each strand of the DNA could make a new strand exactly like its previous partner, and all that would be needed would be a non-specific zippering mechanism, were essential to accepting the idea that DNA could be the genetic material.

Really cool trick #4 High speed throwing

The idea of a self-replicating molecule was so revolutionary that it had to be tested immediately. The implication of the model was that, every time DNA replicated, one strand would remain intact. There would be two ways to test that hypothesis. The first way was relatively straightforward and would support the hypothesis but was not absolute proof. One could grow cells in the presence of radioactive isotopes and then switch them to non-radioactive media. The next time that the cells divided, the amount of radioactivity per chromosome (seen using a microscope, with appropriate conditions) should drop to half the original, as the two radioactive strands separate, and then remain constant in the next divisions. This result was achieved and was confirmatory, but it involves many assumptions, such as the assumption that the DNA molecule stretches the entire length of the chromosome and is not broken during division, and is difficult to do accurately. In any case, we know that occasionally chromosomes do break during division. The more elegant and conclusive experiment was elegant because it harnessed an original assumption and an unusual but effective technique. Matthew Meselson and Franklin Stahl knew that elements came in different forms, called isotopes. Not all isotopes are radioactive, and they for the most part react chemically in exactly the same manner. However, they differ slightly in weight, so that the densities of the molecules of which they form a part differ. Such is the origin of the term "heavy water," which is a form of water containing the isotope of hydrogen, deuterium, that makes it 11% more dense than ordinary water. One milliliter of heavy water weighs 1.11 grams, rather than the usual 1.00 grams. Such a heavier isotope exists also for nitrogen, with the heavier form about 7% heavier than the most common form. Nitrogen is an important but not the only constituent of DNA, so that the DNA of bacteria raised in heavy nitrogen (obtained from reactors) would be just barely more dense than ordinary DNA. However, there is a means of separating two materials of very similar density. If the DNA is dissolved in a solution consisting of the salts of heavy metals, and the solution is centrifuged at extremely high speeds (of the order of 100,000 x the force of gravity, at which a cube of sugar would weigh approximately 650 pounds), the metal salts will slightly concentrate toward the bottom of the centrifuge tubes, forming a gradient of lighter density at the top and higher density at the bottom. Molecules of DNA will move toward the bottom of the tube, stopping when their density equals that of the salt solution. Meselson and Stahl[41] hypothesized that, if the Watson-Crick interpretation were true, DNA from bacteria grown in heavy nitrogen would be dense, while DNA from ordinary, control bacteria, would be less dense. When bacteria were switched from heavy to light nitrogen, the first division would yield all DNA at an intermediary density (one old strand heavy, one new strand light), the second division would yield half intermediary density (one old strand heavy, one new new strand light) and half light density (one new strand light, one new new strand light), and subsequent divisions would yield increasing amounts of light DNA but a constant amount of intermediary DNA. By forcing the cells to divide in synchrony, they could make the measurements at the appropriate times. And that is exactly what they got: one strand was conserved, more or less indefinitely, and a new strand was built at each replication. It was an elegant and most dramatic confirmation that DNA was replicated by template and therefore could serve as the material from which genes were made. See illustration[42]. There was no further question: DNA was the genetic material. The question became how it could be.

Really cool trick #5: Word puzzles

Francis Crick took a mathematical approach. He based his argument on the observation that genes were known to be linear on the chromosome. This argument was based on the argument that genes known to be on the same chromosome in *Drosophila* would occasionally separate from each other, at different but constant frequencies. Beginning with T. H. Morgan, who identified the locations of genes on the giant chromosomes of these insects by changes that accompanied mutations (Fig. 8.1), geneticists had come to understand that the frequency with which they separated reflected their distance from each other. If one has a string of 1001 beads and pulls it until the beads separate, the chance of the break occurring between any two adjacent beads is 1/1000, and the probability that any two beads will be separated by the break is proportional to the number of beads that separate them. The same was valid for chromosomes. The probability that any two genes would separate was proportional to the distance between them, and the genes could even be mapped on a chromosome. With the rise of bacterial genetics, it was even possible to find different mutations within a single gene and map them, proving that the genes themselves were linear. Crick reasoned that, if the genes themselves were linear, one form of coding might be by counting the number of bases. A given number of bases might code for an amino acid. If such a code existed, then the minimum number of bases to code for one amino acid would be three, because of this logic: There are 20 different amino acids. If each base codes for a single amino acid, four bases would code for only four amino acids.

A	T	G	C

Fig. 8.8. One base codes for an amino acid: 4 possibilities

If each two bases codes for a single amino acid, then with four different bases one has 16 possible combinations, or 16 possible amino acids.

AA	AT	AG	AC
TA	TT	TG	TC
GA	GT	GG	GC
CA	CT	CG	CG

Fig. 8.9 Two bases code for one amino acid: 16 combinations

Close, but no cigar. If each three bases codes for a single amino acid, then with four different bases one has 64 possible combinations. This would be more than enough, and would work if the sequence was linear and non-overlapping, as if it could be counted by reading bases in groups of three with no punctuation.

AAA	AAT	AAG	AAC	ATA	ATT	ATG	ATC
AGA	AGT	AGG	AGC	ACA	ACT	ACG	ACC
TAA	TAT	TAG	TAC	TTA	TTT	TTG	TTC
TGA	TGT	TGG	TGC	TCA	TCT	TCG	TCC
GAA	GAT	GAG	GAC	GTA	GTT	GTG	GTC
GGA	GGT	GGG	GGC	GCA	GCT	GCG	GCC
CAA	CAT	CAG	CAC	CTA	CTT	CTG	CTC
CGA	CGT	CGG	CGC	CCA	CCT	CCG	CCC

Fig. 8.10 Three bases code for one amino acid: 64 combinations

THEBADBOYFEDTHEFATCATANDDOGTHEBIGREDBUG

(THE BAD BOY FED THE FAT CAT AND DOG THE BIG RED BUG)

Could he prove that this was the case? Fortunately, some chemicals cause mutations by getting tangled in the DNA helix, causing it to bend and allowing an extra base to be incorporated when the DNA replicates.

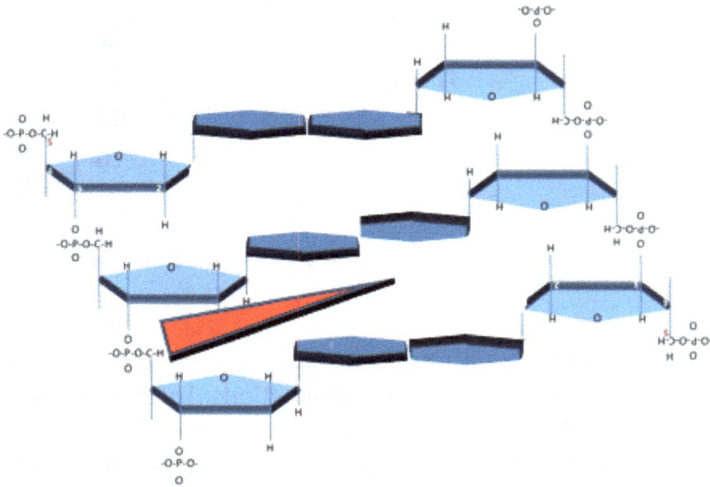

Fig. 8.11. A spatial representation of DNA, indicating how a drug (actinomycin D) causes DNA to add a base. Light blue: Deoxyribose chain; dark blue: bases; red: actinomycin D. Part of the molecule is flat and intercalates (slips between) the bases. When the DNA replicates, the drug clings to the DNA and the new strand, interpreting the drug as another base, adds an extraneous base to the sequence.

Radiation, on the other hand, damages a base and causes it to be skipped when DNA replicates. Crick therefore proposed a test of his hypothesis. The DNA presumably carried the code for (information to build) an enzyme, which needed a specific sequence of amino acids to work. If the DNA was read in a 3.3.3.3. sequence, then an addition should produce either nonsense or a completely erroneous structure, and the enzyme would not work:

THEREDDOGBITTHEFATBOYBUTTHEBIGGALRANOUTANDGOTTHEBAT.

THE RED DOG BIT THE FAT BOY BUT THE BIG GAL RAN OUT AND GOT THE BAT.

Addition:

THE RED dDO GBI TTH EFA TBO YBU TTH EBI GGA LRA NOU TAN DGO TTH EBA T

Similarly, if a base was missing, the sequence would fail.

THEREDDOGBITTHEFATBOYBUTTHEBIGGALRANOUTANDGOTTHEBAT

THE RED DOG BIT TEF ATB OYB UTT HEB IGG ALR ANO UTA NDG OTT HEB AT

However, if he could create a bacterium that contained both mutations, then the reading would come back into sequence and the enzyme might be defective but might still work:

THEREdDDOGBITTHEFATBOYBUTTHEBIGGALRANOUTANDGOTTHEBAT

THE REd DDO GBI TTE FAT BOY BUT THE BIG GAL RAN OUT AND GOT THE BAT

This is exactly what he got.

Crick would not have been able to conduct this experiment had there not been a prior birth of what was to become the field of molecular biology. This birth was achieved by a really remarkable bit of, quite literally, kitchen chemistry. The issue was to study the functional and structural relationship of genes to each other. Since T.H. Morgan's studies of the *Drosophila* chromosome, it was known that genes were aligned on the chromosome, but what determined when a gene was active, or how two genes interacted that, for instance, affected the same color, was not. However, *Drosophila* was not a good candidate for analyzing genes. Although it was very convenient, such analysis would require measuring properties in thousands or tens of thousands of individual flies, a very laborious procedure. The time between generations, two weeks, was short but still long in terms of collecting different mutations. Finally, the fact that it was diploid meant that a new mutation, if it was recessive, would be hidden by the dominant form of the gene until two carriers mated, a situation that logarithmically expanded the number of animals needed. Since most mutations were likely to be lack of an ability to make something, most would be recessive. Fungi such as the black bread mold *Neurospora* bred much faster, but they too required considerable manual labor and had very few characteristics to measure, such as the color of the mold. Nevertheless, George Beadle and Edward Tatum had used them to hypothesize that each gene represented the information to make a single enzyme. (We now know that an appropriate definition is a bit more complex than this, but this definition is useful for our purposes.)

Really cool trick #6: Kitchen chemistry and the birth of molecular biology

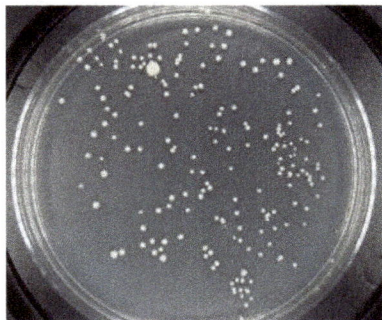

Figure 8.12. Bacterial plaques assay. These are colonies of bacteria, each of which started from a single bacterium that started to grow approximately 24 hours previously. Thus each bacterium in each spot is genetically identical to others in the spot. There are hundreds of thousands of bacteria in each spot (clone). Given unlimited access to nutrients and otherwise ideal growing conditions, one bacterium could theoretically produce 5,000,000,000,000,000,000,000 bacteria in 24 hours.

Using mutants to study mechanisms was obviously a good idea—we can find out if this bulb is the blinking bulb in the Christmas string of lights by replacing it with a different bulb—but hoping to find the right mutation was not the way to go. Herman Muller improved the situation by showing that X-rays could cause mutations, and then producing a lot of them in fruit flies, but even fruit flies take two weeks to mature and reproduce. George Beadle and Edward Tatum had introduced the common bread mold *Neurospora* as an experimental organism. It reproduced much more rapidly and for far less cost. Using it, they had established that, as a general rule, one gene produced one enzyme (see BTW1[7]), but *Neurospora* is, after all, a mold. It doesn't do much. There are different colors of the mold, which provided the basis for Beadle and Tatum's hypothesis, but it is difficult otherwise to study much about the means by which genes work. Bacteria would be much better organisms with which to study genes and mutations, but they had two apparently insuperable problems. First, since they were haploid (only a single copy of each gene) and most of what one could measure would be the ability or lack of ability to make something, you could identify a new mutant only by the fact that it would die if you deprived it of a key nutrient—in which case you just lost the mutant you had identified. Second, bacteria were not known to recombine sexually. That meant that you might find one mutation, for instance, a form that could not make the amino acid arginine, and another mutation, a form that could not make the amino acid tryptophan, but you could not get the two mutations together in the same bacterium to study how they interacted. Tatum pointed out these problems to a new graduate student, Joshua Lederberg, noting that, if one could find a way around these problems, bacteria, with a generation time of 20 minutes and the ability to produce millions of new individuals overnight, would be wonderful for the study of genetics.

Lederberg first found a way not to lose new mutations. What he did was maddeningly simple, in the sense that brilliant experiments usually lead to a "Why didn't I think of that?" response. Simply put, he made a rubber stamp. What he really did, as the story is told, is that he borrowed a piece of velvet from his wife. As a rubber stamp picks up ink and then imparts it in a defined pattern onto successive pieces of paper, Lederberg reasoned that he could make stamp copies of bacterial colonies. If you leave a petri dish containing nutrient agar exposed to the air, you soon see round droplets on it. These are colonies of bacteria, each a clone of thousands of identical bacteria produced by one bacterium that has fallen on the plate (Fig. 8.12). You can of course get a similar result by washing a solution containing a few bacteria of known type onto the dish. What Lederberg did was to grow bacteria in a complete medium, so that all the bacteria would grow, then to use a version of a rubber stamp, which would pick up some but not all bacteria from each colony, and stamp that pattern onto a dish that lacked a particular nutrient. The beauty of this arrangement was that he could identify a mutant by its inability to grow on the second dish, but that the mutant—rather, its identical brothers and sisters—were still available on the first dish. He could go back and pick them up, confident that he had a colony of mutant bacteria. This trick was useful enough, but what it led to was very important. To study how genes interacted, he took one mutant that could not make the amino acid arginine (arg-) and another mutant that could not make tryptophan (trp-). He plated about 10,000,000 bacteria onto Petri dishes containing media that lacked these ingredients. Normally, one might expect "back-mutations" to occur about once in 1,000,000 duplications, so he would expect to find approximately 10 colonies that could grow in each dish. Now he plated both types together onto a dish that lacked both arginine and tryptophan. He expected approximately 10 bacteria that could make arginine but not tryptophan (arg+, trp-) to back-mutate to be able to make tryptophan and grow, and 10 bacteria that could make tryptophan but not arginine (arg-, trp+) to back-mutate to make arginine and grow. Instead, he got about 100 times more bacteria that could survive on the plate lacking both arginine and tryptophan.

This could have occurred because each bacterium made and secreted the amino acid that the other one lacked, but it was possible to test for secretion of these amino acids, and in any case the surviving bacteria (reproduced again through replica plating) should not have been able to survive when plated in isolation onto new dishes. He ruled out both of these possibilities. He concluded: Each mutated bacterium in isolation could back-mutate approximately one time in 1,000,000. When the two mutated forms were combined, the frequency of recombination increased to 100 in 1,000,000 or 1 in 10,000. The most plausible interpretation was that the bacteria had sexually recombined, so that a new chromosome containing both the ability to make arginine and the ability to make tryptophan was created. He thus established that bacteria had sex, and by doing so launched the era of molecular biology, which is based on our ability to rearrange the chromosomes of bacteria and other microorganisms. It is also, incidentally, the manner in which bacteria acquire resistance to antibiotics or even several antibiotics. For instance, if a bacterium is resistant to penicillin and crosses with a bacterium that is resistant to streptomycin, one can produce a new bacterial chromosome that carries resistance to both penicillin and streptomycin. The poly-resistant "superbugs" and "flesh-eating bugs" (which continue to do damage because they are not stopped by antibiotics) were created either by sexual recombination or by the bacteria picking up and incorporating a foreign piece of DNA in a manner similar to sexual recombination, as in the Avery- MacLeod-McCarty "transforming factor".

Really cool trick #7: Talking to molecules

Replica plating allowed Crick to create the kinds of combinations of mutants that he used to establish that the genetic code was linear and that it most likely consisted of three bases in DNA carrying the information to encode one amino acid in a protein, but the code—which three bases indicated which amino acid—was not yet known. The story of how the code was translated is another wonderful adventure story, but one that requires a little bit of explanation as to how proteins are actually made. The information concerning the sequence of amino acids in a protein resides in the DNA, but the DNA remains in the nucleus of a cell, whereas the protein is made in the cytoplasm, outside of the nucleus. The information gets to the site of manufacture by an elaborate apparatus. The coding strand of the DNA is copied as it normally is when DNA is synthesized, but the copy is a slightly different molecule called RNA. This RNA differs from DNA in that the sugar is ribose, not deoxyribose (hence RNA, not DNA) and in that one of its bases is uridine, not thymidine. So a copy would be as follows:

Normal DNA replication		RNA copy of DNA	
(Red is the new strand)			
DNA	DNA	DNA	RNA
A	T	A	U
G	C	G	C
C	G	C	G
T	A	T	A

Figure 8.13. DNA replication and synthesis of mRNA. In DNA replication, a complementary base is added to a new strand across from each base of the old strand. When RNA is synthesized, the same process occurs, except that the sugar is ribose, not deoxyribose, and uridine is used rather than thymidine.

This RNA, the messenger RNA or mRNA as we described earlier, leaves the nucleus. The protein-synthesizing machinery is rather bulky and not very moveable, so the mRNA travels to it. This machinery, called the ribosome, acts like a big press that simply stamps out parts or, in this case, links them together. It can link any amino acid to any other. The template that it uses to choose which amino acids to attach is the mRNA, which ratchets along the ribosome complex as each codon (unit coding a single amino acid) is read. The amino acids are brought to the ribosome by stockboys called transfer RNAs[43] or tRNA. Each tRNA carries a single amino acid at one end and, at its other end, a sequence of three RNA bases reflecting the code for that amino acid. So, if the DNA reads ATG, the messenger RNA made from it reads UAC, and the tRNA that will bind to that spot of the message will contain the sequence AUG. The tRNA in this instance would carry the amino acid tyrosine. tRNA is the true bilingual dictionary, in the sense that at one end it speaks nucleic-ese (AUG) and at the other it speaks protein-ese (tyrosine).

Before it was known that the mRNA existed, it was understood that the DNA had to be transcribed (copied into) RNA before being translated into protein. At that time, the best "A for effort" attempt to identify the code was by Martinas Ycas, a mathematically-oriented biologist at Cornell University. He pointed out that one way to identify the code was to find a protein with a highly unusual composition and to look at the RNA in the cell that made that protein. There was a caterpillar in Africa that made a cocoon with a silk that consisted of only two amino acids, glycine and alanine. He went to Africa, collected the caterpillars, and analyzed the RNA from their silk glands. Unfortunately, it proved to be essentially identical to RNA from any other gland or animal. He concluded that, if RNA carried the code, then the coding RNA had to be less than 2% of the total RNA. He was right, but it was disappointing. A few years later it had become clear that a small fraction of the RNA, now known as mRNA, carried the code to the cytoplasm. There commenced a big effort to identify each type of RNA and determine its function.

Marshall Nirenberg and Heinrich Matthei were trying the same thing, and in particular were examining how enzymes known as ribonucleases degraded the different types of RNA. To conduct their experiments they made an artificial RNA that they could label and follow, which in this case was an RNA that consisted only of uridine, UUUUUUU…, and which they called polyU. (They used this because U is found only in RNA, not DNA, and it was easy to make a radioactive form.) They planned to follow the functionality of the polyU by adding it to a "protein-synthesizing kit," in which a mixture of all the necessary energy sources and different types of RNA could produce a net yield of protein. This worked because the ribosomes and tRNAs from all sorts of plants and animals are extremely similar or identical and will work together, as long as an mRNA is added to provide instructions. The polyU was their artificial mRNA, and they could measure its functional stability by seeing how well it could serve to make protein. To their surprise and disappointment, the intact polyU when added to the kit produced a precipitate. It is difficult to work with precipitates, and the appearance of the precipitate usually indicates that something is wrong: one of the salts is too concentrated, the pH (level of acidity) is wrong, or one of the reagents is bad. However, Nirenberg and Matthei did not discard their solutions in disgust and with an obscenity. They decided to see what the precipitate was. It turned out to be a long string of phenylalanines, insoluble because the single amino acid phenylalanine is itself very poorly soluble in water.

They realized that the protein synthesizing kit had worked but had made an artificial protein. More importantly, they had talked to a molecule and it had replied. They said, "UUU, UUU, UUU, UUU" and the kit had replied, "Oh, I see. Phenylalanine, phenylalanine, phenylalanine, phenylalanine". They had the first codon, or word in the bilingual dictionary.

To say that the news was electrifying would be the gentle way of phrasing it. Nirenberg announced their finding at a meeting of biochemists in Moscow in August, 1962. On his way back to the US, he was invited to stop in England and France to tell about their results. He returned to the US in September and was invited shortly thereafter to speak at Harvard. At that meeting, someone unknown to us (I was a graduate student at the time) got up and, with a laboratory notebook in his hand, said, "Since you made the announcement, we have identified several more codons," and proceeded to read from his notebook. We learned subsequently that Severo Ochoa, at Columbia University in New York, had been in Moscow and had telephoned back to his laboratory, an expensive and complex thing to do at that time. His laboratory had immediately begun to look for other codons, and the unknown person was an emissary who had flown to Boston to present the results when Nirenberg spoke. The next year or two consisted of a feverous and often acrimonious effort to identify all of the codons, and they were finally all identified. The activity of the apparently extra 44 codons (64 possibilities, 20 amino acids) was worked out. Some do not code for any amino acid and serve as "stop" codons. Where this code appears in the mRNA, there is no tRNA to match it, and the protein chain is terminated. This is a means of telling a protein where to end. The start of a protein is always a methionine, coded for by the mRNA AUG (DNA: TAC). Other codes are degenerate, in that more than one triplet codes for the same amino acid. For instance, UCA, UCU, UCG, and UCC all code for the amino acid serine. Whether this serves any purpose is still under investigation.

The argument could be and was quickly tested. For one thing, it was known that the cause of sickle cell anemia, in which hemoglobin precipitates under certain conditions, was caused by a change in a single amino acid from the soluble glutamine to the insoluble valine, which made the mutated form of hemoglobin less soluble. The DNA was searched for the mutation—sequencing, described below (p. 119), had become possible—and, sure enough, the conversion of a single base, A, to T, produced the amino acid change, as predicted by the described codons. Other researchers with considerable effort engineered an artificial tRNA that carried one amino acid but carried the code for another amino acid. (Such mutations are not found in nature, and are very hard to produce, because for obvious reasons they are lethal.) Again, using a protein-synthesizing kit, this artificial tRNA could be demonstrated to insert the wrong amino acid into a protein. That is, it bound to the mRNA at the position dictated by its anticodon, but put the amino acid that it carried into that position.

The fact that the code is universal is really quite amazing, and is one of the strongest arguments to support the hypothesis that all life today is descended from a single source. Consider this: First, the entire structure by which the information how to make proteins is encoded in DNA and proteins are made by transcribing the code into mRNA, which is then delivered to ribosomes, to which are brought tRNAs carrying the specific amino acids for the protein, is conserved. With essentially minor differences in the structures of the ribosomes, the mechanism is the same from bacteria to humans, and even viruses use the machinery of the cells in which they reside. Next, although there is no physical constraint that requires a tRNA carrying a specific anticodon to bind only a specific amino acid, each tRNA is uniquely designed to carry the amino acid for which the anticodon is designed. Again, with very few exceptions, message from bacteria or viruses can be read correctly and accurately by mammalian ribososomes, and any combination of mRNA, ribosomes, and tRNAs from any source—bacteria, plants, worms, insects, mammals including humans—will function and can be used in the laboratory. This is even more startling than if Columbus' translator had faced the Arawak or Carib Indians and listened to them speaking pure Castilian 15th C Spanish. The point is that, so far as we know, there are no physical constraints either to design the protein synthesis machinery in only the form that we see; nor are there physical constraints to force the tRNAs to structure the codes as they do. The fact that our entire world has settled on one arbitrary mechanism argues very strongly that the original design was so effective that all living things on this planet descended from that original design. The same argument would apply, less abstractly, to language. Since language is arbitrary—there is nothing about the sounds "dog," "chien," "cane," "perro," "Hund," "sag" (to use only a few Indo-European languages) that attaches specifically or uniquely to the creature that is a common household pet—if everyone on the planet spoke only English, and only with an American New England 21st C accent, we would have to consider that there was a connection among all the inhabitants of the planet. This is not to commit to the argument that God created life as described in Genesis, or that any other god created life as described in any other holy book. For those who wish to believe so, the unitary descent of living creatures is consistent with such an origin, but the fact does not imply a Creator or a necessary plan. Nor does it imply that there was only one origin to life. It is quite possible that as chemical mixtures acquired the characteristics of consuming energy to construct their own materials and acquiring the ability to reproduce themselves (the characteristics of life) there were other solutions, by mechanisms and routes totally open to the imagination. However, until we find evidence to the contrary, it would appear that other forms, if they existed, were less successful; and that they were out-competed by the form of life that we have, and disappeared too early in the history of the earth or at too primitive a stage, too amorphous a form, for us to recognize today.

Thanks to the universality of our genetic makeup and function, we can conduct many of the types of study of mechanism and rearrangement of biology that we today call genetic engineering.

However, to do so, we require many new tricks and inventions to provide the tools to perform these miracles. The story of the rise of genetic engineering is a saga in its own right and is described next.

~~~~~

Chapter 9: Genetic engineering

Figure 9.1. A small fragment of a chromosome, spread out on a water interface. The base of the chromosome is the black material at the bottom. The DNA strands, still containing a lot of protein, extend outward from it. A few strands of DNA have been teased free at the top. Paulson and Laemmli 1977[44]; All of the fine strands are DNA. This article was published in Cell, Vol 12 J.R. Paulson and U.K. Laemmli, The structure of histone-depleted metaphase chromosomes, Pages 817–828, Copyright Elsevier (1977). Credits: From J.R. Paulson and U.K. Laemmli, 1977. Cell

There is a LOT of DNA in a cell. We have 48 chromosomes, containing enough DNA to make 1,500,000 genes, though we actually have only 20–25,000 genes. The other 98.4% of the DNA was at first considered to be useless or "junk DNA". Today we know that in some cases that DNA is the DNA of viruses that have inserted themselves into our genes, sometimes being used to our purposes. In other cases the DNA appears to be an inadvertent duplication of another gene that now has a mutation that makes it inactive or useless. In many other cases the DNA immediately surrounding a coding gene (one that produces a protein product) is used to determine in which tissues and at what times in development or life the genes are active. For instance, the major muscle proteins are made only in muscle, and the proteins made in heart muscle, skeletal muscle, and the muscle that lines our digestive tract are each different.

Somewhat disappointingly, in spite of our sophistication, frogs have even more DNA than we do. The question is, how do we make any sense of it? To do this, we have to be able to handle a small, known piece of DNA so that we can determine the sequence of its bases, and from there begin to understand how that sequence determines its activity and what it does. Luckily, in the world of biology, almost everything that can be done has been done by some organism in a specific situation, and we can learn from these organisms. For instance, we can learn how to cut up DNA into small, known pieces.

Really cool trick #8: Tom Sawyer and the fence to whitewash

The most perfect parasite does not harm its host, and may even help the host. After all, if the host thrives, the parasite gets a free ride. If the parasite kills the host, then it either has eaten its own seed corn or it has to find a new host. We carry many parasites, most of which do no harm or even, as gut bacteria or gut flora, help us to digest our food. The nastiest diseases are those that have recently jumped from another animal to us and have not yet evolved a means of getting along with us without killing us. Viruses do the same thing, and many infect bacteria. There is a problem with this. Many bacteria have one chromosome, which is circular. What would in our bodies be one end of the chromosome joins to the other end, forming an intact circle. Sometimes the bacterium needs to break that loop, to untangle the newly forming chromosome from the older chromosome when it divides. The virus likewise can go along for the ride most efficiently and with least harm to the bacterium if, rather than floating freely in the cytoplasm, it is integrated into the bacterial chromosome. However, there are potential problems here. If the virus cuts into the chromosome at any random point, it risks disrupting a gene that is vital to the bacterium (the bacterial chromosome has much less non-coding DNA than an animal or plant chromosome), thus killing the host and defeating the virus' goal of survival. Also, viruses often opt to cut themselves loose from the host DNA, reproduce in large numbers, killing the host, and take their chances on finding a new host. They need to cut themselves free rather cleanly. To do this, viruses and some bacteria have developed restriction endonucleases. "Endonuclease" means that the enzyme will cut DNA in the middle of a strand, rather than chewing it up from one end. "Restriction" means that it will not cut the DNA anywhere, but only at very specific sequences. These specific sequences are a series of four to six bases, for instance GAATTC.

The sequence is interesting on two counts. First, the probability of starting with "G" is ¼, since there are four types of bases. The probability that the next base will be "A" is also ¼, and so forth. The probability of getting all six bases aligned in this sequence is ¼ x ¼ x ¼ x ¼ x ¼ x ¼, or about 1 in 4000. The DNA of fruit flies has about 122 million bases and 14,000 genes, and the DNA of humans has about 3 billion bases and 20 to 25,000 genes. If the sequence of bases is essentially random, the DNA of flies could be cut into 30,000 pieces and the DNA of humans, into 750,000 separate pieces. This would mean that, if we had a book that contained as punctuation only periods (a stop codon) and no spaces or paragraphs, we could break it into units if we separated it every time we encountered the structure "hesaid" or "chapter". We have a tool for getting manageable fragments of DNA, which will be useful for many purposes, as is described later (p. 124).

Another interesting property is that the sequence recognized by the endonucleases, of which there are many, is always a palindrome. Note what the two paired strands of the DNA described would be:

GAATTC

CTTAAG

If you read the upper sequence from left to right, it is the same as the lower sequence read right to left, as in ABLE WAS I ERE I SAW ELBA or MADAM I'M ADAM. This is OK, since the two DNA strands are arranged head to tail. (The linkages are asymmetric, in that the linkage on the left side of a base is different from the linkage on its right side, so that the strands can be seen as having head and tail ends.) The restriction endonuclease does not really care which strand it is attacking, so this particular endonuclease will cut the upper strand between the G and the first A, and the lower strand between the G and the first A to the left, separating the DNA into two pieces with single strands dangling over the ends. These single strands will become important shortly. The cut pieces would look like this (the sequence "atgc" is added in lower case to indicate any sequence to the left and right of the target cutting point):

atgcatgcatgcatgcGAATTCatgcatgcatgcatgc

atgcatgcatgcatgcCTTAAGatgcatgcatgcatgc

becomes

atgcatgcatgcatgcG   AATTCatgcatgcatgcatgc

atgcatgcatgcatgcCTTAA   Gatgcatgcatgcatgc

The consequences are profound. First, we have small segments of DNA, which we can hope to analyze. Second, these small segments are likely to differ among species and even among individuals, giving us a means of measuring many aspects of the evolution of DNA and of organisms, and even of identifying individuals, as explained further below (Cool tricks #9 and #10). Third, any piece of DNA cut by this enzyme will have the same overhanging ends, giving us the <u>possibility of reattaching any DNA to any other DNA</u>. This is the means by which the bacterium or virus recloses the DNA loop, and it is the basis for our recombining genes in the process called genetic engineering, producing recombinant DNA. To perform all of this laboratory magic, we need to have a few more tricks up our sleeve.

First, for the most interesting purposes, we need to be able to take a miniscule amount of DNA and produce more, so that we can measure it. For instance, we would like to take a swab from a crime scene, or the DNA remaining in a Neanderthal bone, and analyze it, or in a hospital we may have to determine the organism that is causing an infection from a very small sample, or we would like to know if the cells in small biopsy carry a mutation diagnostic of cancer. Second, we need to be able to separate the segments of DNA, which we will do by size. Third, we need to be able to determine the sequences of the pieces to conduct the analyses for the purposes in the first point. Fourth, we need to stitch the fragments of DNA back together if we wish to change the genetics of any organism for any research or commercial purpose.

Really cool trick #9: Intellectual tourism (observing the crazies)

For many purposes you want to get your hands on a vanishingly small amount of DNA. For research purposes you would like to know what genes are in a baby mouse that, if it has the right mutation, you want to raise and breed. You would like to get that DNA from a snippet of its tail, so that the mouse will survive the assay and grow to found a colony of mice with the mutation. A forensic scientist is re-examining a 10-year old rape case, for which the only evidence is a cervical swab taken from the victim, containing a few sperm on a dried microscope.[45] You learn that you can get some 40,000 year old DNA from the marrow of a Neanderthal bone and want to determine how closely Neanderthals were related to modern humans. You have a blood sample from a patient who apparently has a viral disease but is not getting better. You want to know what the virus is, but you cannot grow it in culture. You are trying to trace the origin and spread of a new disease by identifying changes that have occurred and from that constructing a family tree. For these and many other situations you can get some DNA, but it is in vanishingly small amounts, much too small to analyze. How can you get more?

We know how DNA replicates, but it is not easy. You know how difficult it is to untangle a bunch of coiled springs or a garden hose that has been rolled up and gotten tangled. We have the same problem with DNA. The double helix must be unwound, so that new strands can be built on the old, now single strands. This takes a complex series of enzymes to accomplish this, ultimately using one very complex enzyme called DNA polymerase which, leaning on the old strand and taking its cue from the old strand as a template, attaches the appropriate bases opposite to the template and onto the growing new strand, assembled in order from one end to the other. The enzyme complex itself does a sort of group hug to the DNA, actually breaking it so that the complex can force its way in and unwind the two strands so that the assembly can take place. It then repairs the breaks on the two strands. Biological systems obviously do this efficiently, but it is very costly and difficult in the laboratory. The most difficult and expensive part is the unwinding step, carried out by a complex called the "unwindase". It is relatively easy to unwind DNA into single strands. If you heat DNA to a temperature of about 140° F, it spontaneously unwinds. However, at this temperature the rest of the complex, the "polymerase" or assembly part (it produces a polymer or long string of bases from the monomers or free individual bases) cooks like an egg, destroying the activity of the enzyme. If you unwind the DNA at high temperature and let it cool, it will reassemble into a double helix. In fact, one of the early ways of determining how similar different species were to each other was to heat a mixture of DNAs from two species, unwinding the DNA, and letting it cool, in which case the DNA strands would randomly try to associate with each other. It was possible to judge how effectively the hybrid strands held together by the temperature at which they adhered or separated, giving an estimate of the similarity of the DNAs. Kary Mullis was contemplating the problem of the cooked polymerase when the thought crossed his mind: In hot springs, such as those found in Yellowstone Park, live bacteria. Different species of bacteria thrive at different temperatures, creating the spectacular concentric rainbow colors of the springs (Fig 9.2).

Figure 9.2. Chromatic Spring, Yellowstone National Park. The hottest water in the geyser is in the center, and it cools as it leaves the center. Each color represents a different species or ecosystem of microorganisms that can live at that temperature. *Thermophilus aquaticus* lives in the white areas just away from the center.

Some of the bacteria survive at temperatures astonishingly close to boiling. If the bacteria live and reproduce at these temperatures, by definition their DNA polymerase survives and is functional at that temperature. So he went to Yellowstone, collected the bacterial species *Thermophilus aquaticus* (literally, the heat- loving creature found in water), and extracted its polymerase. He demonstrated that he could heat DNA from other organisms to unwind it and use the bacterial polymerase to synthesize new strands of DNA at high temperature, at which the substrate DNA remained unwound. This was the P of the PCR reaction, referring to the now commercially available Taq (for *T. aquaticus*) polymerase. Running this cycle once doubles the original amount of DNA, which may not be very helpful if a doubled amount is still vanishingly small. However, if you repeat the process, the total amount of DNA goes to 4X the original amount. That's where the "CR" of "PCR" comes in. "CR" means "chain reaction," and the full name is "Polymerase Chain Reaction". If you repeat the cycle 10 times, you increase the amount of DNA 1000 fold. If you repeat the cycle 30 times, you increase the amount of DNA one billion times. This is the normal practice for PCR. The cycle is run up to 30 times, producing enough DNA to study for any of the several purposes described above. Of course, it is not quite as easy as television crime shows present. Without extreme care and cleanliness, in collecting the sample and in processing it, you are likely to get a billion-fold amplification of the hot dog you ate for lunch or something that you sneezed.

Really cool trick #10: Racing DNAs

Now that we can cut DNA into small pieces so that we can handle it and amplify DNA so that we can analyze it, we now need to be able to separate it into pieces with which we can work. It still does no good to work with a mixture of everything (scientists tend to use the German word "Gemisch"). We will do this by setting up a DNA race. The principle is very straightforward and had been known for some time. DNA molecules, like proteins, tend to have a small negative charge under the conditions in which they are held in solution, and therefore they will move in an electric field toward the anode, or positive pole. For instance, if you touched a wire to the outside or bottom of a standard "D" battery, put that wire in a salt solution containing DNA, and led another wire from the solution to the center pole of the battery, the DNA would move to the wire touching the center pole. We do essentially this in a process called electrophoresis, though it works much better at 60 to 100 V. The migration, however, will be relatively fast and uncoordinated. We have to discipline the migration so that we can use it. We do that by trapping and slowing down the DNA as it moves. We can do this with gels, as is shown in the animation[46].

Gels (think Jello® or the slimes that children love) consist of a meshwork of fibers immobilizing water trapped within the fibers, like a sponge or paper towel. Because the fibers (gelatin, which is boiled gristle, for Jello) are much more tightly intermingled than those in a sponge, the water is more tightly held and less easily dislodged, but you can actually squeeze water out of gelatin if you try. It is a less messy proposition if you put it into a centrifuge. The gel will pack to the bottom, leaving the extruded water on top. By controlling the concentration of the gelatin or, in the laboratory, a plastic or carbohydrate that will polymerize into fibers, one can produce a gel in which the pores between the fibers are of known dimensions, for our purposes of approximately the same size as the molecules we are interested in studying. Now, when we put the DNA at the cathodic (negative) end of the gel and turn on the current, a situation ensues that would be equivalent to a group of portly adults trying to go through subway turnstiles at the same time as a group of grade-schoolers[47]. As you might guess, the children scamper through quite easily, while the portly adults have more trouble squeezing past the barriers. For molecules, the smaller ones move quite quickly through the gel, which to them is mostly water, while the larger ones bang into and get caught up in the fibers. The result is that after the electrophoresis has run for a while, the molecules are separated according to size. This works for proteins or for DNA. If we have a means of coloring or labeling these macromolecules, we can, at least initially, identify them by their size. Although size is a crude criterion, it often is the first time that something of interest is identified. We therefore have names for important proteins like p70, meaning "unknown protein of 70 kiloDaltons in size".

There is one immediately useful function of being able to separate DNA in size. Because we can use restriction endonucleases, we can detect specific types of very small variations in DNA. For instance, consider the following sentence as a code:

> In the city we find many interesting things and many plants and animals, but we do not usually find many wolves or deer.

Now assume that our verbal restriction endonuclease cuts up this sentence everywhere that it finds the word "many":

> In the city we find many/ interesting things and many/ plants and animals, but we do not usually find many/ wolves or deer.

We get fragments of 24, 28, 52, and >16 characters. Suppose now that there are very slight changes in the sequence, either within the recognized sequence "many" or outside of it.

> In the city we find many/ interesting things and mUny plants and animals, but we do not usually find many/ wolves or deer.

Now we have fragments of 24, 80, and >16 characters.

> In the city we find many/ interesting things and many/ **TYPES OF** plants and animals, but we do not usually find many/ wolves or deer.

Now we have fragments of 24, 28, 61, and > 16 characters.

> In the city we find many/ interesting things and many/ ~~plants and~~ animals, but we do not usually find many/ wolves or deer.

Now we have fragments of 24, 28, 41, and >16 characters.

These fragments can be used, by size distribution, to distinguish the DNA of one person from that of another, for many forensic and other studies. Of course, in the former situation these are crude measurements, by size only. Otherwise, the sequence of a fragment can be read directly and the difference of a single nucleotide noted. The point is that we can find such variations quite routinely, often in single bases located where they do not have any significant effect on the biology or health of the individual. These are the famous SNPs or Single

Nucleotide Polymorphisms If we could actually determine the sequence of DNA, we could tell, for instance, how closely a chimpanzee is related to us, how closely chimps are related to other mammals, and we could determine many other relationships. On the valid assumption that the DNAs of two creatures are more similar the more closely related they are to each other, we can examine from which animals whales descended (this turns out to be hippopotamus-like animals) and where vertebrates came from (our closest relatives are the—very exciting and intellectually stimulating—starfish and sea urchins). We can trace human migrations, determining which people are most similar to those who originally left Africa and the relationship of the geographical races in the world. Even more than that, we can determine that the rate in which these minor variations in DNA arise is relatively constant, so that we can use the extent of difference to estimate when, for instance, the people of the new world became isolated from the people of the old world, meaning that they had settled in the Americas, or when the lines of evolution leading to chimpanzees split from the line leading to humans. But to do this, and to understand how genes work, we need to be able to sequence the DNA.

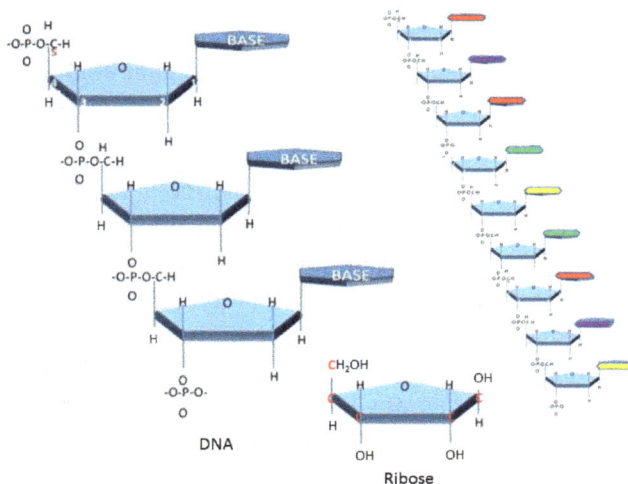

Fig. 9.3 Bottom center: The physical configuration of ribose, a 5 carbon (C) sugar. Each carbon is bonded to an oxygen (O) or hydroxyl (OH) group. Carbons are numbered 1-5 starting at the right and proceeding clockwise. Upper left: The general structure of DNA. In deoxyribose, the oxygen on the second carbon (2) is absent. In DNA a "base" (one of four possible nucleosides) is attached to each carbon 1; phosphates are attached to carbons 3 and 5, and successive deoxyriboses are linked through these phosphates. Upper right: In a complete DNA molecule, the sequence of the four bases (here represented by different colors) designates the sequence of amino acids that the DNA will encode or otherwise specifies the activity of the DNA.

Really cool trick #11: Watch the machine do the work.

To understand how DNA sequencing is done, we need to have a bit better picture of the structure of DNA, most particularly the backbone on which it is built. The backbone is built from a modification of the sugar ribose which, like all sugars, has an oxygen atom attached to each carbon atom and, like many sugars, forms a ring, such as in Fig. 9.3, left. Each corner except the one with an "O" in the center represents a carbon atom. The carbons are counted clockwise from the right. Deoxyribose, from which DNA is made, is similar, but it lacks one oxygen, specifically in carbon position 2. The backbone of DNA is formed by the oxygen at position 5, now converted to a phosphate, linked to the oxygen at position 3 of the next ribose. Thus the chain is built. When DNA is synthesized, each sugar is added successively to the next sugar, linking it between the third carbon of one deoxyribose and the fifth carbon of the next. Now the question is how to make sense of the whole structure. You cannot just take a necklace consisting of four types of beads, break it up into its constituent beads, and hope to put it back together in the order that it was. You have to take it apart bead by bead, recording the order in which they appear. It takes two steps to do this. The first is a mechanism to take the chain apart bead by bead or, in this case, base by base. It involves a chemical dead end called dideoxyribose. This sugar analog lacks an oxygen not only at the 2 position but also at the 3 position. Thus the building DNA chain cannot be extended by attaching another deoxyribose to it.

Now let's see how this will work. Let us prepare a solution in which a new strand of DNA can be synthesized, using DNA polymerase to replicate the old strand. Into this solution, which contains all four nucleotides (bases attached to deoxyriboses), A, T, G, and C, we add a dideoxynucleotide, for instance ddA. Now, every time that the sequence calls for an A to be added, some of the molecules will add the dideoxy A and thus terminate, while others will add the normal deoxy A and continue. However, at the next point at which an A should be added, some of the "survivors" will now add the dideoxy A and terminate at that point. We will therefore get a series of short chains, each one representing where the chain stopped at an A. For instance, if the sequence is GATTCGAGCTAGGCCA, we will get the following pieces:

GddA

GATTCGddA

GATTCGAGCTddA

GATTCGAGCTAGGCCddA

and, reading from the bottom (anode) end of the electrophoresis, we will get sequences 2, 7, 11, and 16 bases long, indicating that there was an A (T in the original strand) at positions 2, 7, 11, and 16. If we run this experiment separately for each of the four bases, we will get a ladder reading the sequence off in succession.

This is not the whole process. We still have to identify them. If we make each of the four dideoxy bases radioactive, so that they will expose X-ray film, we can locate each chain and, counting up from the bottom (the shortest chain) we can read the sequence of DNA (Fig. 9.4; also see YouTube description[48]). In practice, we can read a few hundred bases at a time and, by overlapping sequences, extend the known sequence quite a bit. This technique however has its limitations. The electrophoresis is often not as clean as one might wish, and it is difficult to tell whether one band is lower than the next, particularly since four separate experiments have to be run. Thus, this technique, the Maxam-Gilbert sequencing technique, has given way to a more elegant version. The more elegant version takes advantage of the fact that the 3 position is useless for binding the next ribose, and instead we attach to that position a fluorescent molecule. Now, instead of looking for radioactivity, which poses some dangers and expensive precautions, we can look for fluorescence to locate the bands. Better yet, we can find four different fluorescent molecules, so that each dideoxy base fluoresces a different color. In the illustration, A is green, T is red, G is yellow, and C is blue. Now we can mix all four dideoxy bases together in the same mixture and run one gel, since we can now tell all the bases apart. Even better, we don't even have to stop the reaction to read the gel. If we run the electrophoresis off the end of the gel, we can set up a sensor to read each fluorescent molecule as it leaves the gel, and automate the process so that the machine happily ticks off the sequence as the fluorescent bases appear. This is the automated sequencing apparatus that has produced so many headlines involving sequencing the human genome, identifying mutations causing particular diseases, tracing our relationship to Neanderthals, and many other wonders.

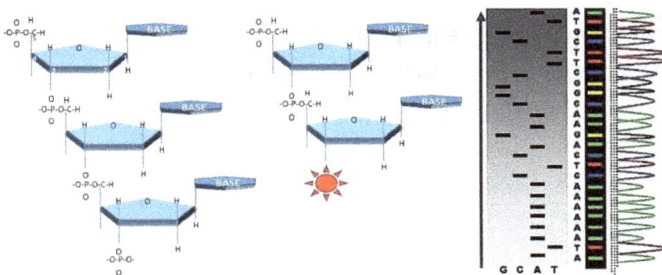

Figure 9.4. Left: The structure of DNA and the principle of DNA sequencing. DNA is built by a chain of sugars (deoxyribose) that are attached through phosphates from the #3 carbon in one sugar to the #5 carbon in the next sugar, while the bases are attached to the #1 carbon (see numbers in topmost sugar on left). The #2 carbon lacks an oxygen, giving rise to the name "deoxyribose". Middle: The linkage at carbon #3 is replaced by a fluorescent molecule. The chain cannot be extended beyond this point, and reproduction produces a shortened chain. This chain can be collected based on its length, and the fluor identified. Since the base of the nucleotide to which it was attached is known, this specific nucleotide and its position are known, allowing one to assemble a list of the sequence of nucleotides, colloquially called bases, in the DNA. Right: An idealized pattern of a DNA

sequence as determined by the Maxxam-Gilbert technique, labeling four samples of DNA individually with dideoxy G, C, A, and T, and reading the gel from the bottom (smallest piece) upward, or by using dideoxyfluors, running the samples through a liquid column, and allowing a machine to detect the fluorescence signals.

Really cool trick #12: Mr. Potato Head

We cannot leave the topic of genetic engineering without considering the final (for now—the future holds many surprises in store) and eponymous trick of genetic engineering: fitting the parts together in a different order. Consider the fragments generated by our restriction endonuclease above:

atgcatgcatgcatgcG   AATTCatgcatgcatgcatgc

atgcatgcatgcatgcCTTAA   Gatgcatgcatgcatgc

Note that the enzyme will cut any DNA containing that sequence, and will produce the same overhangs everywhere it cuts. If we can only get the two fragments back close together, the A's and T's will interact almost as well as before, and the chain will hold together. In fact, our bodies have prepared for such an accidental breakage, as well as the breakages that occur when new DNA is synthesized, and we have an enzyme, DNA ligase, that will link the G back to the A at the breakpoint. This is a major means by which we repair the damage caused by the sun, by cosmic rays, by intrinsic radioactivity, and by other mechanisms. The DNA ligase recognizes only the breakpoint, and is not at all finicky about what lies to the left or right of the breakpoint. Suppose, then, that we were to add to the mixture not the end that we broke off but an end produced by the same endonuclease but from a different DNA. This different DNA could be a good copy of a gene mutated in the original DNA, a marker that allows us to tell when the gene is active, something that we deliberately insert into an animal to see how it works, like a gene for Alzheimer's Disease, or something completely foreign, such as a gene from a bacterium that produces a product that inhibits the formation of ice crystals. Such genes have been inserted into plants to make them resistant to frost. The point is that we can attach any piece of DNA to any other piece of DNA cut with the same enzyme:

atgcatgcatgcatgcGAATTCatgcatgcatgcatgc

atgcatgcatgcatgcCTTAAGatgcatgcatgcatgc

and

ggggccccggggccccGAATTCggggccccggggcccc

ccccggggccccggggCTTAAGccccggggccccgggg

can be cut into

atgcatgcatgcatgcG   AATTCatgcatgcatgcatgc

atgcatgcatgcatgcCTTAA   Gatgcatgcatgcatgc

and

ggggccccggggccccG   AATTCggggccccggggcccc
ccccggggccccggggCTTAA   Gccccggggccccgggg

and these fragments can be combined into

atgcatgcatgcatgcGAATTCggggccccggggcccc
atgcatgcatgcatgcCTTAAGccccggggccccgggg

What this means is that, for experimental purposes, we can attach a marker ("flag") to a promoter to see when and where a gene is activated, or we can attach a structural gene to a promoter that we can turn on or off at will. For commercial purposes, we can insert into a plant a gene that will help the plant resist frost or infection, or we can make other modifications. (We have been doing this, less efficiently, for centuries: No orange, apple, ear of corn, cat, or dog resembles the wild type from which it was derived; the famous horticulturist Luther Burbank took no precautions with any of the many crosses that he made. BTW1[7]). In medicine, we can create antibodies or drugs that address specific problems or, eventually, restore functions lost by disease or inheritance. This is genetic engineering. The technology is available to create, for instance, extremely dangerous microorganisms, though elaborate safeguards are in place when previously dangerous creatures such as the 1918 influenza virus are studied. No sane scientist intends to do otherwise.

So this Brave New World of genetic engineering is upon us. Though technologically advanced, it does not differ in kind from what humans have always done. Pekinese dogs and pugs were not spontaneous developments from their wolf-like ancestors, nor were the popular bug-eyed, four-tailed goldfish. In fact, most of these creatures are deformed and, like dachshunds, great danes, and whippets, have substantial problems. Nor do most of the foods we eat look like their wild ancestors. We prefer them to be bigger and more colorful, with higher sugar content, and less likely to spoil. Genetic engineering is simply a means of making the selection more efficient.

Nevertheless, in the argument that science is an onion (see above) learning so much about the structure of DNA has only brought more questions. Now that we know what genes are and how they work, we learn that much of what makes us (and all of life) so interesting is not so much the particular genes and their protein products, but when and where these genes are turned on and off, and how those controls work. After all, in terms of genes and protein products, we are nearly identical to chimpanzees. The timing and placement of when genes are active and how active they are, from those that make hair to those that determine the position of our larynx (so that we can speak) to those that determine the growth and maturation of the brain, make all the difference. Opening this layer of the onion of questions is what makes biology continuously interesting, and makes every successive day the most exciting day to participate in the science enterprise.

~~~~~

Chapter 10. Conclusions--What is science, and where do we go from here?

The scientific mode of thinking can be described as a type of philosophy—a mode and structure of analysis. Its basis is the assumption that an analytical interpretation of the evidence of the senses is the best means of understanding our world. It does not rely heavily on the sensual or emotional side of human experience (passion) as a guide to interpretation, because it has no means of weighing, experimenting with, or falsifying the meaning of passion, trances, or other emotive experiences, and it considers to be irrelevant, evidence based on immeasurable factors such as faith, communication with the dead, extrasensory perception, telekinesis, or other considerations beyond human experience. Note that the operative words are "considers irrelevant": the scientific approach does not reject out of hand such evidence; rather, the scientist states that he or she cannot test and therefore evaluate such evidence and therefore cannot incorporate it into a logic of the workings of the world. For science is about the mechanics of the world, how the world is put together and how it functions. Science does not consider why the world is here.

The rules of evidence:

Humans have always used the rules and logic of evidence, even in the most adamantly faith-based procedures known to mankind. What were miracles but evidence of the existence of a superior being? And trials by ordeal in all faiths were an effort to establish evidence. They ranged from the African ordeal bean, in which someone accused of an offense to society was made to eat a bean containing a deadly neuropoison, to torture of the accused in a court of Puritans, or the Inquisition. In each case the survival of the accused was evidence of innocence, and death was evidence of guilt. The rule was still evidence, but the logic included assumptions of untestable forces ranging from the power of God to unknown forms of energy. As long as they remain untestable, they are beyond the reach of science and the scientific approach. They may exist—before the existence of the microscope, bacteria were inconceivable, and before the discovery of radioactivity, the idea that a rock could explode and release enormous amounts of energy was unimaginable. Scientists simply say that we know of no forms of energy and no mechanisms by which ghosts, for example, could exist and come to haunt the earth. We can attempt to detect their existence, by setting up numerous detection devices and, above all, attempting to reproduce the conditions in which they appear. If we fail in spite of our best efforts to capture an unequivocal and measurable sign of their existence, they remain an unproven hypothesis, currently falsified by evidence supporting the opposing hypothesis that ghosts do not exist. The evidence supporting the hypothesis that ghosts do not exist is weak, since it consists entirely of negative evidence: the ghost was not recorded by a camera, motion sensor, heat sensor, magnetic field detector (such as a metal detector), microphone, or any of the numerous other means we have of detecting distortions in the environment. As we were surprised to learn that bats use sonar, whales and birds may sense magnetism, and insects and birds can recognize star patterns, we can still be surprised to discover a new modality to be sensed. Any well-planned and executed experiment that detected a ghost would in a single step overturn the hypothesis that ghosts do not exist, but we would then have to move to the next step of logic—how do they exist? What is their source of energy? Of what are they composed? Science merely tells us where to place our money in a bet, and in this case the best wager is that ghosts do not exist.

Talk to boring people:

A major source of difficulty is the perception that scientists are arrogant, peremptory, and intolerant of other views. This should not be the case. At worst, we ask that others understand the framework of our logic, that we accept as "real" or "relevant" (within the scope of our expertise) only arguments that can be tested, and that we have rules for the evaluation of evidence. Beyond that, we can learn much from others. We all need to listen.

James Watson in 2007 wrote a book that was entitled *Avoid Boring People*. While he has many reasons to make such an argument, and many would agree with him, sometimes one can learn a great deal from those who are not considered to have the highest academic potential. Certainly some of the experiences that I have had have colored my viewpoint as to what is possible with the human mind. I do not have answers any better than those who make it their profession to study how to educate people, but I have learned why it is a mistake to be too judgmental.

Math skills are not necessarily displayed in academics.

When I was in grammar school, one of the poorest-performing students was the son of a teacher in the school. This teacher quietly approached me and asked me if I could help her son, so that at least he would not fail arithmetic. I tried, and it was a struggle. He had trouble with the logic of long division, and the rudiments of algebraic skills ("word problems") were torture. However, of course, we chatted during the lessons, and we both liked baseball. It took me some years to appreciate, but I finally began to think about those conversations. If you asked him, "What is Larry Doby's batting average?" he could answer, "Yesterday it was .315. If he gets three hits out of five at-bats today, it will go to .317." Presumably, he was approximating rather than doing the calculations, but nevertheless, it was a sense of mathematics and the rules of numbers. If our assignments had been problems concerning the statistics of baseball rather than cars going down the street or dividing cookies between Jane and Bill, he most likely would have done much better. Education (and the result thereof) is often all about catching a child's interest and holding it.

The ability to think abstractly is the ability to turn the abstract into the concrete.

Three memories come back to me:

- One summer I was working on a ground crew at a local college. One of my co-workers was my age (teenager) who had grown up in an orphanage and was at the time beginning to move into the world of work. His education did not appear to be very profound. One day, while we were having lunch, someone was bouncing a basketball a few hundred yards away. Sound, of course, travels much more slowly than light, and the sound of the basketball bouncing arrived considerably after we saw it bounce. I made a comment about it, including some remark about the speed of light (perhaps "That shows that the speed of sound is slower than the speed of light"). He asked what I meant, and really could not get his mind around the concept that light would have a speed. "I see it. That's when it happens." Since the delay in sound was obvious, he could deal with the idea that sound took time to reach him, but not that light could take time to reach him.

- As a graduate student, I had freshman advisees. One young lady was taking a physics-oriented "science for non-scientists" distribution course, and was having trouble. I tried to help. I would set up the calculation, a simple arithmetic equation, in apples and oranges. No problem. I would say, "I'm getting tired of writing 'apples' and 'oranges'. I'll just write 'a' and 'o'." No problem. I would say, "You know, 'o' looks like '0'. Let's do it the way others do. Instead of 'a' and 'o', I will use 'v' for velocity and 'a' for acceleration." Complete stop. She could no longer do the problem. Nor did she have any idea how to begin. And she was panicked. Conclusion: it was not the ability to calculate, but the ability to use an equation to address a concept (velocity) rather than a thing (apple) that daunted her. Again an issue of abstraction—to process something intangible as if it were real. It can be taught, but how do you teach it? (Follow-up: she survived the course and went on to a strong career in politics. She was known nationally.)

- I found it easy to describe the function of an oscilloscope (an instrument used in physiology) by relating it to old-style television sets, the principle of which had been a subject in high-school physics. Similarly, I could describe several physiological functions by describing them in terms generally reserved for the old radio vacuum tubes. However, as transistors took over electronics, I found it difficult to understand and explain coherently how they worked, and my students likewise failed to grasp my allusions to televisions or vacuum tubes. Conclusion: What I had learned in high school and prior to high school was part of my real world, what I took for granted and upon which I could expand. Other analytical or logical abstractions were more purely intellectual exercises, and substantially more difficult to absorb.

- I finally learned a trick in teaching, when I did not have models. I would tell the class, "You see, when I talk about molecules, I keep moving my hands. They have real shapes, and they can bend and twist. I can picture them and imagine their movement." And then I would elaborate as much as necessary, trying to get students to let the abstraction "molecule" grow in their minds until it was a tangible, palpable thing. Today we have 3-D rotating models available on the web and in PowerPoint, and they are best used frequently to help students bridge the gap between the visible and the unseen.

What I had learned: The ability to think abstractly is the ability to turn the abstract into the concrete. A huge step in education is to get the student spontaneously to take a concept and to turn it into a thing, a real, physical structure that can be examined and manipulated. It is a single, stone-wall barrier that can be crumbled. As biologists, we have the relatively easy task of trying to get students to imagine viruses (which can, in a sense, be "seen" in electron microscopes) while physicists must communicate atoms, sub-atomic particles, and electron orbits, and mathematicians must conjure something real from a property that can be only imagined.

Precision in language.

One summer, in a factory, one of my co-workers routinely chose me as a lunch-time conversational companion. (In retrospect, he may have been shunned by everyone else, and I was available. I don't know.) The problem was, his conversation would usually open with (expurgated but verbatim), "That f*'n son of a b* of a f*'n son of a b* f*ed me." It would take me ten minutes of non-committal "yeah"'s and grunts until I could gather enough information to guess what he was talking about. Conclusion: Precision and clarity of language—"to crystallize and transfix the moment"—underlies precision and clarity of thought.

It is an absolute requirement for scientists, and a source of much aggravation and confusion when scientists and non-scientists interact.

Many talents and many peoples create science.

When I entered science as a student, American science still had the patrician tone of its English ancestor. Most faculty came from Harvard or Stanford, the two leading schools of the few Ph.D. producing institutions of a prior generation, and experiments were done in a cool, gentlemanly, almost leisurely, ratiocination-based analysis of possibilities before an experiment was done. A Boston Brahmin accent overhung everything. What I first noticed was the difference in style of the seminars given by the weekly invited speakers. These latter were some of the best and brightest in the country, many heading toward future Nobel and other prizes. But they talked fast, most commonly with New York accents; they waved their hands a lot; they cracked jokes about their work; and they were very excited about it. The theme seemed to be, "You mean, I can play in this mud pie my whole life and get paid to do it?" It was what I came to describe as peasant science, the "let's play with it, see how it works, suck it and see" style, born not of the aristocracy but of peasant immigrants, willing to get their hands dirty and to take chances, a pragmatic, earthy style. It was enormously productive and produced the efflorescence of American science that dominated the second half of the 20th C. What we learn is that there are many ways and styles to build good science. As that generation of scientists of essentially European ancestry gives way to a new generation, which will in the US include far more Asian, Hispanic, and Black participants, one waits and looks for what the talents and experiences of their societies will bring to the enterprise. It should be different, and it is very likely to be good.

Unfettered curiosity and persistence pay off

This argument can be told with some anecdotes about frogs:

The idea of cloning an organism is quite ancient but in biological terms began with Joachim Hämmerling's experiments with *Acetabularia* in the 1920's through 1950's. *Acetabularia* is a genus of umbrella-shaped single-celled marine algae that comprises several species, each of which has a different shape of umbrella, with the nucleus of the cell sitting at the base of the "pole" of the umbrella. Hämmerling sectioned the algal cells into three: base (containing the nucleus), stalk (pole), and cap (umbrella) and then put the pieces together in different orders, including mixing pieces from different species. He found that the base was required to maintain the regeneration of the cap. When he attached the cap of one species to the base of another, eventually the shape of the cap converted to that of the species that had contributed the base. With further refinements, he demonstrated that the type of cap was determined by the nucleus of the cell, the first such demonstration for eukaryotic cells (those that have a true nucleus and organelles, as opposed to bacteria).

These of course were algae, and Hämmerling changed only the shape from that of one species to another. However, in 1952 Robert Briggs and Thomas King did something more startling and more noteworthy to the lay observer: They cloned a frog. Specifically, frogs' eggs are fertilized externally, and one can perform the task manually by adding sperm to ripe eggs in water. In the process of fertilization, the penetration of the sperm "activates" the egg so that it initiates development, and the nucleus of the egg rises to the surface, where it will meet and fuse with the nucleus of the incoming sperm. Pricking the egg with a fine needle simulates the penetration of the sperm and can activate the egg, and, using a dissecting microscope, one can see the female nucleus rising to the surface and use a tiny pipette to suck or flick it out of the egg. One therefore has an activated egg that has no nuclei. It can undergo a few divisions but ultimately

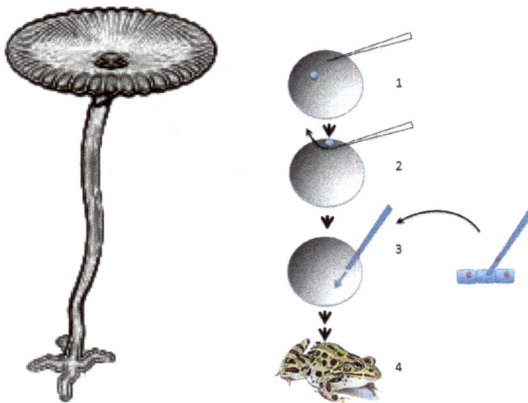

will fail (Fig. 10.1).

Figure 10.1. Left: An Acetabularia plant. The plant can be as big as 10 cm (4 in) tall, but is generally smaller. The nucleus resides in the "root" structure at the bottom. Right: The Briggs and King experiment. 1. The unfertilized egg is pricked with a needle. This emulates the penetration of the sperm and initiates development. 2. As part of this activation, the nucleus of the egg rises to the surface, where it would normally meet the nucleus of the sperm. It can be seen under a microscope and can be flicked out of the egg by using another needle. 3. Using a specially prepared glass micropipette with a sharp edge, a cell from an embryo in later development is punctured and the nucleus is sucked into the micropipette and then injected into the activated, enucleated recipient egg. 4. Under favorable circumstances, the egg will continue to develop and turn into a tadpole that will metamorphose into a normal frog, but one that bears the characteristics of the donor nucleus.

Briggs and King sucked out nuclei from other embryos, 2, 4, 8, 16, 32, or 64 cells (that is, eggs that have advanced 1-6 cell divisions), and injected them into recipient, activated but enucleated, eggs. Many of these transplanted nuclei survived and took over the development, eventually producing healthy tadpoles that grew and metamorphosed into frogs that then matured into fertile adults. This was the first true cloning of a vertebrate. However, there was a limitation. The first ten or so divisions of an egg are called cleavage. These divisions are more-or-less uninterrupted, with the cells doing little more than dividing. We now know that they function on information (messenger RNA)[49] placed in the egg by the mother. After the 10th division, cell division slows down and becomes asynchronous. We now know that at this time the embryos start to use their own messenger RNA and begin the process of differentiating into distinct tissues and organs. Briggs and King found that, while they could get tadpoles from nuclei collected before the transition, all tadpoles formed from nuclei after the transition were extremely abnormal and died before or just after hatching. It appeared that, although the nuclei contained the same DNA, the transition created a limit to their potential.

Marie A. DiBeradino provided the first clue as to what the limit might be. Carefully studying the aborted tadpoles, she recognized that they had grossly abnormal chromosomes. Pointing out that the rate of cell division slowed considerably at the transition, she suggested that the limitation might be mechanical, a failure of the transplanted nuclei to get themselves ready to divide as fast as the cytoplasm was moving. Following this argument, she demonstrated that nuclei taken from a frog tumor, in which cells were dividing rapidly, could produce tadpoles. Thus the limitation was a technical limitation, not necessarily a change in character of the nucleus. By 1975, a Swiss group had initiated normal embryonic development from the nucleus taken from an antibody-producing lymphocyte.

John Gurdon had picked up on these ideas and set about to test the limits of the nucleus. He worked in England, where the common American leopard frog (the preferred research animal in the US) was much harder to get. Instead, he chose to work on the South African clawed frog, which was available in England. As a laboratory animal, it had several advantages over the leopard frog. It could be easily bred and kept in aquaria, whereas the leopard frog generally did not thrive in captivity and had to be collected from the wild; its breeding was not as seasonal as that of the leopard frog; and one could collect eggs several times from one female, whereas the inability to feed and rear leopard frogs meant that one could collect only the eggs developed before the frogs were captured in the fall. Finally, there were at least two mutations known in clawed frogs, the number of nucleoli in the nucleus (which could be identified before the embryos differentiated) and an albino mutation.

Gurdon assiduously transplanted thousands of nuclei from various organs and finally got some surviving tadpoles, mostly from fairly large rapidly dividing cells, such as from intestinal epithelium, from which he could extract the nuclei without mechanical damage. The percent success rate however was very low, and one could argue that he occasionally failed to destroy the nucleus of the host egg, thereby allowing the egg's development. To counter that argument, Gurdon used animals with differing numbers of nucleoli as host egg and donor nucleus, and he transplanted nuclei between albino and normally colored animals. In one famous picture, he photographed the brown donor of the eggs, the albino parents of the tadpoles from which he took the nuclei, and thirty-six albino frogs that he had raised from the transplants. True cloning of vertebrate embryos from differentiated cells had been achieved. With subsequent refinements, it is now possible to achieve much higher success, and to clone cells with nuclei from much more complex cells.

--

The African clawed frog proved interesting for another reason. When the frogs were first used to study development, a common and tolerated procedure was to make a small incision in its belly, collect some mature eggs, make a rudimentary effort to suture the wound, and put the frog back into the rearing tank. Since the frogs tend to secrete a lot of slime and they don't always eat the food offered to them, the tanks were pretty dirty. (Since then, animal welfare laws mandate much more proper conditions for maintenance and egg collection.) I remember asking, the first time I saw the early procedure, "Don't they get infected?" and getting the response, "Naw, they're very hardy." And we went about our various tasks. In the sense that scientists should always be curious about everything, several years later another scientist, Michael Zasloff, asked the same question and decided to look further. He found that the frogs produced in their skins an antibiotic, which when purified proved to define an entirely new class of antibiotics, now called magainins (from the Hebrew for 'shield'). They have not proved to be as commercially valuable as the microbial antibiotics, but their mechanism of action provides a focus for the design of new antibiotics, and the discovery helped launch the idea that antimicrobial peptides were widely dispersed in the animal and plant kingdoms. We can conclude that (1) if you live in dirty water, it's a good idea to bring along your own antibiotic and (2) one should ask questions about everything.

--

On that note, for the incorrigibly lazy among us (me), frogs may have a future contribution to make. Anyone who has tried to hold a struggling frog knows that its hind legs are quite strong. They are not exceptionally strong, as the force generated is a function of both the cross-sectional size of the muscle and the leverage that it can attain, but they are good muscles. Consider this, though: frogs don't do yoga, cardio workouts, isometric training, or strength training. They sit all day, pretty motionless. How do they maintain their muscle tone? Good question.

Science is amoral

It is also very important to remember that morality is a human trait but that science is amoral. By "amoral" we mean that science does not have morals, that it is neutral to morality. Science is not "immoral," or against human codes of morality. It is amoral, in the sense that the value of any human action or judgment is a human decision for which science can provide evidence but not interpretation. A scientist can state when the genome of a new human being is created and at what point the nervous system is developed to the level at which we can presume that an infant feels pain or has a thought, but the value of that information, meaning whether or not the state or a church assumes interest and responsibility for that life, is a value judgment made by societies, and the conclusion has varied from society to society and throughout history, often much more harshly than we would decide today. Nature gives no comfort to the concept of sanctity of life: for most creatures, and for humans for most of our existence, infancy is or was a hecatomb, with predators waiting for baby turtles or fish eggs to hatch, and the predators often including members of the same species or even the parents themselves. Many human societies have considered the lives of infants of little value. Some societies (Sparta) even selected, at the age of one year, the "better" of a pair of twins and discarded the other. These are facts to which specialists can attest. What we make of it today is our decision as a society; scientists may vote, but they do not decide the morality of what we choose. Likewise, we can provide evidence that evolution has occurred and our analysis of this evidence can inform our predictions as to what will happen if we raise the carbon dioxide or methane level in the atmosphere (global warming) or what we will lose if we destroy great ecosystems such as the tropical rainforests. We can likewise interpret how genes will and will not spread in our population, or calculate how many people this planet can hold. But we cannot make decisions for a human society.

Science is a matter of aesthetics

To repeat an argument made previously: Science is simply a style of inquiry like any other, and choosing to be a scientist is an esthetic decision. Everyone can ask, "Why is a sparrow brown?" And there are many valid answers: "Because God made it brown." "So that it can hide from predators." "Because female sparrows are attracted to male sparrows with the purest brown color". "Because it has melanocytes and other pigment-containing cells in its feathers." "Because during its early development melanocytes migrated to the regions where the feather follicles would grow." "Because in the very early stages of development, some cells laid down chemical tracks for the melanocytes to follow." "Because melanocytes contain melanin granules, which contain melanin, a brown-to-black pigment." "Because melanin absorbs light across the spectrum of visible light."

"Because melanin is derived from tyrosine, and its double-bond ring structure allows it to absorb the energy of photons in the range of visible light." And so on, until we get deep into the theory of how the positions of electrons in atoms affect the ability of molecules to capture light, and finally to the theory of atomic structure. At each level of inquiry, some people will say, "That answer satisfies me. The rest is too dry and dull." Those who stick with the inquiry into the technical details are the scientists. Others follow the inquiry into different directions and may become theologians or poets, but all are simply trying to makes sense of the world. Which direction we follow is simply a matter of esthetics.

As long as humans have inquiring minds and are permitted to inquire, science will advance and will not stop. Humans will continue to evolve. (Within the last 10,000 years or less, some populations have evolved at least twice the ability to drink milk as adults, at least twice the ability to breathe and live at high elevation, and resistance to several animal-borne diseases. We are currently mixing previously separated geographical races and are presumably heading toward a future when most humans are tan colored. As a whole, humans are moving toward having fewer children. Depending on social or economic factors, some segments of our societies may favor the procreation of specific types by designating specific aspects of appearance or behavior to be "hot", and thus affect other directions of evolution.) The society must assess its own values, and in this endeavor all participants have a say. Sometimes societies make very bad decisions, and sometimes they make excellent ones. The role of the scientist is to tell us how it works and therefore to predict the consequences of specific actions. Hopefully you, the citizenry, will be sufficiently well-informed to understand the importance of evidence, logic, and falsification, and you will evaluate the data, or at least demand that the data be presented in a form intelligible to you, and make moral and compassionate decisions on that basis. If you can do this, then we as scientists have succeeded in our mission.

To come full circle...

Who are "we"? Well, if you have a job in which you teach science, or if you work in a laboratory in which you have any level of say in what goes on, or you do the equivalent on a field trip or in front of a computer, then you probably meet at least the U.S. Internal Revenue Service's definition of being a scientist. On the other hand, are you curious? Do you wonder how things work? Do you ask questions "because they are there..."? Then, intellectually, we are brothers. If you have trouble with the abstractions, tell me and ask me how to make them more concrete. All I ask is that you tolerate my request for precision, and that you understand my tolerance of ambiguity.[50]

~~~~

About the author:

Richard A Lockshin was born in Ohio and received his undergraduate and graduate degrees from Harvard. He taught at the University of Rochester School of Medicine and Dentistry and later at St. John's University in New York, and is currently Professor Emeritus at St. John's. As a research scientist he is known for his studies of programmed cell death or apoptosis, now a major research topic, a field of which he is considered to be a founder. He has well over one hundred research publications, including several technical books in the field. He resides with his wife on Long Island, New York.

Discover other titles by Richard A Lockshin at Smashwords.com:

Born This Way: How Science is Done and Practiced, by Richard A. Lockshin, due Fall 2013

The Joy of Science: Springer, 2007, available through Amazon

Several other technical books on cell death and on aging, available through Amazon

Connect with Me Online:

mailto:rlockshin@gmail.com

Facebook: http://facebook.com/richard.lockshin

Smashwords: https://www.smashwords.com/profile/view/RichardALockshin

My blog: http://sayingsofthepreachers.net

******

End Notes

---

[1] http://www.businessinsider.com/22-maps-that-show-the-deepest-linguistic-conflicts-in-america-2013-6?op=1

[2] http://en.wikipedia.org/wiki/File:2006-ca-turkey.jpg

[3] http://en.wikipedia.org/wiki/File:Rouge_gorge_familier_-_crop_%28WB_correction%29.jpg

[4] http://en.wikipedia.org/wiki/File:Turdus-migratorius-002.jpg

[5] http://en.wikipedia.org/wiki/File:Meerkat_feb_09.jpg

[6] http://en.wikipedia.org/wiki/File:Cork_Micrographia_Hooke.png

[7] Born This Way: Becoming, Being, and Understanding Scientists. Richard A Lockshin, 2013. Available from Smashwords: https://www.smashwords.com/books/view/317988; Amazon: http://www.amazon.com/Born-This-Way-Understanding-ebook/dp/B00E0N0TQS/ref=sr_1_3?ie=UTF8&qid=1377022405&sr=8-3&keywords=lockshin+richard; Barnes and Noble: http://www.barnesandnoble.com/w/born-this-way-richard-lockshin/1115446743?ean=2940044545991; and iTunes Store (Books>Science & Nature>LifeSciences)

[8]

http://www.fda.gov/BiologicsBloodVaccines/SafetyAvailability/VaccineSafety/ucm096228.htm#thi

[9]

http://www.microbiologytext.com/index.php?module=Book&func=displayarticle&art_id=27

[10] For some references on this material, see the following:
Hatton, John and Plouffe, Paul B, 1997, Science and its ways of knowing. Prentice Hall, Upper Saddle River, NJ.
Carey, Stephen S., 2004, A beginner's guide to scientific method, 3rd Ed., Thomson Wadsworth, Belmont, CA
Piel, Gerard, 2001, The age of science. Basic Books (Perseus Books Group), New York.
Wilson, Edward O. 1995, Naturalist, Warner Books, New York.
Wilson, Edward O, 1996, In search of nature, Island Press/Shearwater Books, Washington, D.C.

[11] http://utenti.quipo.it/colettisb/ipertesto-redi/redi/redi-exp.htm

[12] For a book on the subject, see **The Strange Case of the Broad Street Pump: John Snow and the Mystery of Cholera** by Sandra Hempel (2007); for a website, http://en.wikipedia.org/wiki/1854_Broad_Street_cholera_outbreak

[13] http://en.wikipedia.org/wiki/File:Pellagra_NIH.jpg

[14] http://psycnet.apa.org/index.cfm?fa=search.displayRecord&uid=2004-16485-000

[15] When prophecy fails. Festinger, Leon; Riecken, Henry W.; Schachter, StanleyMinneapolis, MN, US: University of Minnesota Press. (1956). vii 257 pp. doi: 10.1037/10030-000

[16] http://www.youtube.com/watch?v=pds8w7C9FEw

[17] http://youtu.be/6AuYHsMsqv8

[18] *A Conspiracy of Cells: One Woman's Immortal Legacy and the Medical Scandal It Caused (Google eBook) Michael Gold SUNY Press, 1986 - Science - 170 pages*

[19] http://en.wikipedia.org/wiki/Walter_Nelson-Rees

[20] Pollitzer E, 2013. Cell sex matters. Nature 500 (August 2013) pp 23-24.

[21] From: Julius Caesar Cassius, a nobleman, is speaking with his friend, Brutus, and trying to persuade him that, in the best interests of the public, Julius Caesar must be stopped from becoming monarch of Rome. Brutus is aware of Caesar's intentions, and is torn between his love of his friend Caesar and his duty to the republic. Cassius continues by reminding Brutus that Caesar is just a man, not a god, and that they are equal men to Caesar. They were all born equally free, and so why would they suddenly have to bow to another man? On another level this phrase has been interpreted to mean that fate is not what drives men to their decisions and actions, but rather the human condition. http://wiki.answers.com/Q/The_fault_dear_Brutus_lies_not_in_the_stars_but_in_ourselv es#ixzz20LeeRy39

[22] http://en.wikipedia.org/wiki/Albert_Einstein; http://en.wikipedia.org/wiki/Sir_Arthur_Eddington; http://www.pbs.org/wgbh/aso/databank/entries/dp15ei.html

[23] http://www.flickr.com/photos/ral_bornthisway/8656312664/in/photostream

[24] At the height of the polio epidemic in 1952, nearly 60,000 cases with more than 3,000 deaths were reported in the United States alone. However, with widespread vaccination, **wild-type polio**, or polio occurring through natural infection, was eliminated from the United States by 1979 and the Western hemisphere by 1991. http://www.post-polio.org/ir-usa.html - _blank

[25] http://en.wikipedia.org/wiki/Levi-Montalcini,_Rita

[26] This was the second Nobel Prize for something related to cell death. In 1901, Elie Metchnikoff was awarded a Nobel Prize for his early studies of the immune system. He had identified phagocytosis, whereby cells from the immune system eat bacteria and other contaminants of the body, but also dead and dying cells produced by the infection.

[27] http://www.flickr.com/photos/ral_bornthisway/9351033829/

[28] Ultimately, de Duve shared a Nobel Prize with Albert Claude and George Palade for their several discoveries in cell biology, another Nobel Prize related to the field of cell death.

[29] Thus a study of poetry was important to establishing myself as a scientist. Irrelevant here, but in further defense of a liberal education, another factor of some importance to my career was my ability to converse in French and German the first time I went to a scientific

meeting in Europe. It put me in the center of many conversations and contributed strongly to Europeans' remembering me when they were planning future meetings.

[30] http://www.flickr.com/photos/ral_bornthisway/10842395156/

[31] Brenner, Horvitz, and Sulston were awarded the Nobel Prize in 2002 ostensibly for their work in developing the roundworm model, but heavily because the roundworm was the key to understanding programmed cell death.

[32] Actually, one still reads papers in which the authors state, "the cells died by apoptosis," which is equivalent to saying, "he died because his heart stopped". The heart ALWAYS stops when one dies, whether because of loss of blood, chemical imbalance because of infection, kidney failure, or other, lack of oxygen, or any of the myriad ways of irreversibly disrupting the function of the body. Apoptosis is a process, not a cause.

[33] http://www.flickr.com/photos/ral_bornthisway/8642713195/in/photostream

[34] http://www.toptenz.net/wp-content/uploads/2012/01/image003-570x570.png

[35] http://www.flickr.com/photos/ral_bornthisway/8510501659/in/photostream

[36] http://en.wikipedia.org/wiki/File:Polyten_chromosome.jpg

[37]
http://en.wikipedia.org/wiki/Avery%E2%80%93MacLeod%E2%80%93McCarty_experimen t

[38] , http://en.wikipedia.org/wiki/DNA

[39] The X-ray crystalogram produced by Rosalind Franklin that ultimately was inter- preted by James D. Watson and Francis Crick as representing a helical structure of DNA Credits: Franklin R, Gosling RG (1953) "Molecular Configuration in Sodium Thymonucleate". Nature 171: 740–741

[40] http://profiles.nlm.nih.gov/ps/access/SCBBYW.pdf

[41] http://en.wikipedia.org/wiki/Meselson%E2%80%93Stahl_experiment

[42] http://www.flickr.com/photos/ral_bornthisway/9558659222/

[43] http://www.flickr.com/photos/ral_bornthisway/8510501659/in/photostream

[44] http://www.cellimagelibrary.org/images/11023

[45] One of the early cases in which DNA identification was done. It turned out that the alleged perpetrator, who had spent 10 years in jail, was innocent.

[46] http://www.flickr.com/photos/ral_bornthisway/10842395156/

[47] http://www.flickr.com/photos/ral_bornthisway/10842395156/

[48] http://www.youtube.com/watch?v=zYT1s6RJMS0

[49] http://www.flickr.com/photos/ral_bornthisway/8510501659/in/photostream

[50] Further references:

http://www.unesco.org/science/wcs/eng/declaration_e.htm

American Association for the Advancement of Science, 1993, Benchmarks for Science Literacy (Benchmarks for Science Literacy, Project 2061) (Paperback)

Chalmers, Alan 1999,What Is This Thing Called Science: An Assessment of the Nature and Status of Science and Its Methods (Paperback)

Rothman, Milton A. 2003 Science Gap: Dispelling The Myths And Understanding The Reality of Science, Prometheus Books, New York

Tambiah, Stanley J, 1990, Magic, Science and Religion and the Scope of Rationality (Lewis HenryMorgan Lectures) (Paperback) Cambridge University Press, Cambridge, UK

www.ingramcontent.com/pod-product-compliance
Lightning Source LLC
Chambersburg PA
CBHW070733220326
41598CB00024BA/3406